ARGUMENTS COUNTER TO EINSTEIN'S THEORY OF RELATIVITY

AND

THEORY OF COMPARATIVE QUANTUM MECHANICS IN THE MICRO AND MACRO WORLDS

By

I0504374

Mark F. Dennis

ARGUMENTS COUNTER TO EINSTEIN'S THEORY OF RELATIVITY

AND

THEORY OF COMPARATIVE QUANTUM MECHANICS IN THE MICRO AND MACRO WORLDS

By

Mark F. Dennis

Research and Proposal

A proposal to address the questions that arise from Albert Einstein's Theory of Relativity, encompassing both Special Relativity and General Relativity as well as the Grand Unified Theory and Stephen Hawking's Theory of Everything

For questions or feedback on this book, contact Mark at Mdennis99@yahoo.com

ACKNOWLEDGEMENTS

Special thanks to Dr. Thomas Sieland of Embry-Riddle University, for all of his assistance through the years in my physics studies and my graduate papers. Also extended thanks to all my teachers and professors in all the schools I have attended and disciplines I have studied under from seminary theological studies to algebra in high school. It is our teachers who inspire us to learn and understand the world and people around us. And most of all, much thanks to my dear late father, my greatest teacher who was my inspiration in life to learn as much as I could about everything that interested me.

ABSTRACT

Much of modern quantum theory assumes that the rules of physics in this material physical world that we live in are not the same rules of physics in the quantum world. In this physical world, energy is an attribute of an object with mass; in the quantum world, it is taught that energy can exist without mass. In this physical world, an object cannot exist in two places at once; in the quantum world it can, popping in and out of existence. This research hypothesizes counter to these theories in that energy cannot exist without mass much like velocity cannot exist without the bullet or speed without a jet aircraft. Energy must be looked at much like Love, an action verb or transitive verb; it has to have an object and cannot exist by its self. If mass is taken out of the equation, energy must also be null. This hypothesis presumes that the rules of this world and the rules of the quantum world exist under the same controls; it simply may be that we see the quantum world all wrong and apply theories and laws that contradict. A Theory of Everything, where all the rules apply does exist; we just have to learn how this all works and that will take time.

TABLE OF CONTENTS

LIST OF TABLES

LIST OF FIGURES

CHAPTER I

INTRODUCTION

Background of the Problem

Mankind has always marveled about the universe; where did we come from and what are we made of? Questions about the composition of light and how it worked has always intrigued us. In the days of Plato, Aristotle and Socrates, light was considered the fifth element; the "Luminiferous Aether", the stuff stars were made of. It wasn't until Galileo in the early 1600's, two thousand years later, that the question of light's speed was first tested. All that could be determined at that time was that the speed of light was at least 10 times faster than the speed of sound. Not too much later in the late 1600's however, Ole Rømer determined through the study of Jupiter's moons that the speed of light was be 299,792 km/sec or 186,290 miles/sec. This concept became the cornerstone of quantum mechanics today, the study of subatomic matter.

What is energy? How is energy propagated through empty space? Many eastern religions and the new age movement speak of pure energy. Does "Pure Energy" really exist? What is attraction at a distance and how can these forces such as gravity operate across the vast emptiness of space? If the pull of gravity reaches out millions of miles through empty space, what is it that is doing the pulling? How was mass created? What is gravities relationship to mass? What relationship does time have with gravity? Is the universe really accelerating outward or is it collapsing inward? These were questions that drove the genius of Albert Einstein to doubt the scientific theories that were in place in the early twentieth century and develop new ideas of his own. Now, almost 100 years later, Einstein's theories have become the foundation of quantum mechanics and modern theoretical physics. Today in the early twenty-first century, is it not time to question the old theories again?

Can energy exist without mass? If light moves in waves, what is waving? How can transverse waves propagate without a solid medium? Could it be that mass exists simply as energy manifested as standing waves within some sort of rigid substance? How was mass created? What is gravity? What is time and can we go forward or back in time? Do wormholes exist? Do parallel universes exist? Is there an edge of the universe, or a place where time no longer exists? Is the universe really expanding or is it collapsing?

These are all questions that we shall discuss in this book and hopefully, if anything, understand a little more about the universe and the world in which we live, the big and the small. A great deal of effort has been taken to explain the mind and thoughts of Einstein and many other brilliant thinkers throughout history in simple everyday language that the layman and even the young student can understand.

Einstein's Special Theory of Relativity was developed by Einstein after the failure of Michelson and Morley to detect the "Luminiferous Aether." The theory of the Constancy of the Speed of Light grew out of this study and was designed to solve the age-old riddle of why measurements of the speed of light did not show any variation with Earth's movement through space. The speed of light therefore had to be a constant that did not change with the motion of the earth nor any other object no matter what its velocity was. The aether that light was believed to move through also would no longer be necessary, neither would the idea of "stationary space".

As Einstein would put it:

"Examples of this sort, together with the unsuccessful attempts to discover any motion of the earth relatively to the "light medium," suggest that the phenomena of electrodynamics as well as of mechanics possess no properties corresponding to the idea of absolute rest.... Light is always propagated in empty space with a definite velocity of c which is independent of the state of motion of the emitting (or detecting) body. These two postulates suffice for the attainment of a simple and consistent theory of the electrodynamics of moving bodies based on Maxwell's theory for stationary bodies. The introduction of the "Luminiferous Aether" will prove to be superfluous inasmuch as the view here to be developed will not require an "absolute stationary space" provided with special properties, nor assign a velocity vector to a point in space in which electromagnetic processes take place (Einstein, 1905, pg. 1).

Einstein would also say:

"Who would have imagined that this simple law has plunged the conscientiously thoughtful physicist into the greatest of intellectual difficulties?" (Einstein, 1952, pg. 18)

The Special Theory of Relativity would answer the long sought after questions of the past but introduce problems of its own. The standard physical equations for moving objects would no longer hold true. The velocity of a baseball thrown from a moving vehicle would normally be calculated as the vehicles velocity added to the velocity at which the baseball was thrown or $v1 + v2 = v3$. A baseball thrown at 50 miles per hour (mph) from a vehicle driven at 100 mph would be traveling at 150 mph relative to the ground and the person catching the ball. The speed of light is however always the same no matter what the velocity of the emitter or detector. Therefore, using the above equation would change and be written as, $v1 + v2 = v1$ or using the letter c for the speed of light, $c + v1 = c$ and $c - v1 = c$. This contradicts normal thinking, a paradox of sorts, creating some "difficulties". In order to explain this contradiction, Albert Einstein developed formulas to show how that movement of an object warps space and time and that space and time were really one entity, "space-time". He further explained how that acceleration and gravity where truly one in the same with his "Equivalence Principle", and that both also warped space and time.

The velocity of light was fixed; nothing could go faster than the speed of light. Einstein would show through his relativistic mass formula that as an object would accelerate to faster and faster velocities that an objects mass would start to increase exponentially with the velocity as energy would be turned into mass; this theory was founded on the famous formula $E = MC^2$. Mass would be an extension of energy as $M = C^2/E$. Nothing could go faster than light. This then would compound things to a certain degree.

How could electrons and photons then travel at or close to the speed of light? The answer was to make these components massless particles. As we can see, there is an interesting dialogue going on here. In order to answer one problem, a new problem is created. Are these theories truly the way things are or are they really only theories? Do we dare question or challenge them? As we review these theories, let's ask some basic questions of our own. What truly is energy? How is energy propagated through empty space? Is space really empty? What is attraction from a distance (gravity and electrostatic charge) and how can these forces operate across the vast emptiness of space or even from across the room? These were the questions that drove the genius of Albert Einstein to question the scientific theories that were in place in the early twentieth century. Now, almost 100 years later, Einstein's theories have become the cornerstone of quantum mechanics and theoretical physics. Today in the early twenty-first century, is it not time to question these old theories again?

The next questions of this research and what we really want to know is, "is the speed of light attainable? Could someday the speed of light be surpassed? Will this affect time? Can we travel in time? Understanding the questions we are asking takes a basic knowledge of the subjects of Quantum Mechanics, General Relativity and Special Relativity.

In summary, much of modern quantum theory assumes that the rules of physics in this material physical world that we live in are not the same rules of physics that the quantum world exists in. In this world, energy is an attribute of an object with mass; in the quantum world, energy can exist without mass. In this world, an object cannot exist in two places at once; in the quantum world it can, popping in and out of existence.

The foundation of Einstein's Special Relativity is the equation $E=MC^2$ where the energy (E) of a system is equal to the systems mass (M) multiplied by the constant C (Speed of light) squared. This is fundamentally the same equation as momentum (p), the energy of mass (m) in movement (v) or $p = mv$, the only difference is that the velocity is a constant and that it is squared. This is a very simple equation when its basics are understood in that the more mas, the higher the energy and the more velocity, the higher the energy. A major problem with the equation is that energy is converted to mass as velocity increases close to the speed of light making it virtually impossible for an electron or a photon to come near the speed of light; as energy is added, it is converted to mass and gets more and more massive. An exception is given to the photon which qualifies it as a "massless particle" in order to make this equation work and allow the photon to move at light speed, again counter to standard physics.

This research hypothesizes counter to these theories in that energy cannot exist without mass much like velocity cannot exist without the bullet or speed without a jet. Pure energy cannot exist; mass cannot become energy and energy cannot become mass. Energy must be looked at much like Love, an action verb or transitive verb; it has to have an object and cannot exist by its self. Even though energy can be measured quantitatively, it is a character of something; for example a person can have lots of love or lots of energy. Energy is simply manifested as the movement of mass in some way or form; a characteristic of mass. If mass is taken out of the equation, energy must be zero. The hypothesis presumes that the rules of this world and the rules of the quantum world exist under the same controls; it is that we see the quantum world all wrong and apply theories and laws that contradict. A theory of everything, where all the rules apply does exist; we just have to learn how this all works and that will take time.

In his later years, Albert Einstein would struggle to make the quantum world and relativity compatible; a struggle to mathematically unify the gravitational and electromagnetic field theories into on Unified Theory. *"It has been my greatest ambition to resolve the duality of natural laws into unity,"* Einstein said, *"The purpose of my work is to further this simplification and particularly to reduce to one formula the explanation of the field of gravity and of the field of electromagnetism."* Albert Einstein would go on to say that this dilemma had *"plunged the conscientiously thoughtful physicist into the greatest of intellectual difficulties."*

This research will then hypothesize that light waves must move through something, much like waves move through water or air. For a wave to exist something must be waving, a medium or field, at one time called the aether; this medium must also be an elastic-solid that transverse waves can move through since transverse waves cannot move through fluids. Mass therefore must consist of standing waves within this elastic-solid medium giving the illusion of movement while the "pixels" remain solidly in place much like a TV screen. Energy moving through this medium is created by the cooling and coalescing of these particles or confined pressure waves moving through it. This is why the Michelson Morley experiment failed; mass is the medium, manifested by energy waves moving through it.

These hypotheses suggest that variations in pressure, temperature and local time in this medium create the illusive condition called gravity. Here, higher energies away from mass press in on the lower energies of colder, condensed mass (including the fiery stars), and causing mass to attract mass. The "medium" closest to the coalescing mass becomes denser causing light passing through this area to change velocity and refraction occurs much like light passing through glass. "Empty space" therefore varies in density and temperature depending on where that "empty space" is in reference to coalescing mass. Energy therefore moves slower closer to mass due to the mediums increased density and local time will then vary according to the density of that space. The denser the space, the slower time moves ahead in reference to less dense space, time being the change in position of energy moving through the medium, frame by frame. Time also will vary during acceleration and deceleration due to the "principle of equivalence" where gravity and acceleration are indistinguishable; it is not velocity that changes time, it is the acceleration or deceleration to that velocity of the object that changes "local time."

This brings us to some point far away from the center of the "Big Bang" where mass can no longer exist; a place where space is completely void of this medium. Variations in the density of space therefore can produce currents and flow much like the ocean. In the far reaches of the universe where space truly is "empty," energy may move at velocities much higher than our calculated speed of light, counter to the theory that light speed is constant, producing the "wormhole effect." Since mass is the manifestation of energy waves moving through this medium, mass in this region may simply exist, then not exist, and then exist again, "jumping" as it were, far from where it came from. This jump is almost identical to the "quantum jump" the electron encounters as it jumps from one shell to another, existing, then not existing, then existing again.

Small wormholes may also be created here at home, allowing one to "beam" from one place to another. This could be done by "energizing" or heating this medium in close vicinity to a transport vehicle. Energizing the medium would create pressure and temperature differentials in the medium similar to gravitational forces explained earlier. Since the medium that is being energized or heated, consists of so little mass, it may be possible that very little energy is necessary to create this force. It may simply be a matter of frequency, similar to a microwave oven exciting hydrogen the atoms to heat your dinner. This technology may allow the craft to hover or move at extreme velocities without the negative effects of g forces on its occupants, as if "falling" toward its intended destination or possibly "falling up" into the sky.

As energy disperses out from the center where the Big Bang began, the universe begins to cool, slowly collapsing. Redshift seen from galaxies furthest away from us suggests that rather than having an expanding universe where the furthest galaxies are accelerating, the universe actually is slowly collapsing. As energy in our universes core is radiated out into the outer reaches of space, energy is being lost and our core is cooling and shrinking. As we look back into time, an illusion is created that suggests that these galaxies are racing away from us omni directionally; in fact, it could be that we are racing away from them, as if falling into a "black hole." Black holes are all around us, probably the centers of all galaxies. Could it be that black holes are regions where the universe is gone completely cold? Where energy in mass has been completely "wrung out" and the **"Bose Einstein Condensate"** is created? Is this the beginning of the "Big Crunch?"

CHAPTER II

REVIEW OF RELEVANT LITERATURE AND RESEARCH

The Luminiferous Aether

Before the year 1900, most scientists believed that light was propagated through space in what was called the "Luminiferous Aether". They believed that like sound through air, light must travel through something. The belief was that the light-bearing "aether" was a fine gas-like substance made of the smallest of particles.

350 B.C.E - Aristotle Ancient Greeks believed that all things were made of combinations of the four basic substances, fire, water, earth, and air. Among the ancient Greeks were Greek philosophers **Plato** (427-347 BCE) and **Aristotle** (384 BC - 322 BCE), born in the height of ancient Greek civilization. Plato, a student of **Socrates** (469 BC–399 BC), was a philosopher and mathematician who laid the foundations of modern philosophy. Aristotle is believed to be the first person to attempt physics and even gave physics its name – which is the "study of matter" and its motion. Aristotle is one of the central figures of the era and was a student of Plato (Bord, pg. 33).

Aristotle and Plato before him believed that in addition to the four basic elements there was a fifth element which Aristotle would call the aether, meaning "blazing" (Aristotle, De Caelo, Book I, Chapter 2). This aether would be the main constituent of the celestial bodies (Asimov, pg. 6). It would be a substance rigid enough to transmit light at incredibly high speeds while offering no resistance to the movement of the galaxies, stars and planets throughout the universe (Barbour, pg. 128). This was the belief for thousands of years until around the turn of the 20th century when new scientific discoveries came to light, no pun intended, and changed things. Throughout this discussion we will at times call what is sometimes referred to as the aether, a *"field"* or a *"medium"*.

1638 - Galileo - In Aristotle's time it was believed that the speed of light emanating from the celestial bodies was instantaneous or infinite. Although a few thinkers throughout the centuries suggested that light might be finite, it wasn't until Galileo in 1638 that an experiment would be devised to measure the speed of light.

Galileo (1564-1642) lived in Italy during the Renaissance, one of the most productive eras in human history. He was born the same year as Shakespeare and born the same year that Michelangelo died. Galileo gave science the gift of many discoveries in both mechanics and astronomy, such as the sun centered heliocentric model of or solar system, which led to him being placed on trial by the Roman Inquisition. Galileo also discovered the law of falling bodies. Aristotle, hundreds of years earlier had reasoned, erroneously, that the velocity of falling objects were dependent on their weights.

Galileo would later discover that the velocity of two objects of different weights fell at the same rate. Legend has it that this discovery was made from the Leaning Tower of Pisa (Bord, pg. 35). In an effort to understand the velocity of light, Galileo proposed that an experiment over several miles using lanterns, telescopes and shutters be used. By measuring the time it took Galileo to see his assistant's light and knowing the distance of the lamps, Galileo reasoned that he could determine the speed of light. Galileo and an assistant with covered lanterns went to the tops of two hills about one mile apart. The idea was that Galileo first would uncover one lantern, and then as soon as the assistant saw the light of the first lantern, the assistant would uncover his own lantern. The time would be measured from when the first lantern is uncovered until the light was seen from the second lantern. The time calculated was then to be divided by twice the distance between the hilltops, which was a total of two miles. The speed of light should be, **Velocity = 2 miles/ number of seconds.** No noticeable difference was noted during this experiment, thus proving to Galileo that the speed of light was infinite or that light travels at least 10 times faster than sound – far beyond perception of the thinkers of that time (Bartusiak, pg. 24). After Galileo's death this experiment was later performed again at the Accademia del Cimento of Florence in 1667 with no delay noted.

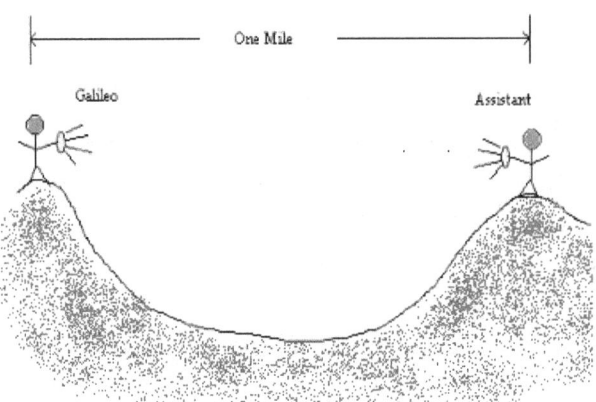

Lantern Experiment

1676 - Ole Rømer (1644-1710) was born two years after Galileo's death and was one of the first scientists to discover the speed of light through studies of the movement of one of Jupiter's moons. Romer was a student of Italian mathematician and astronomer, Giovanni Cassini, who discovered of the Great Red Spot on Jupiter in 1665.

Jupiter takes longer to orbit the sun than the earth does (twelve earth years), so that at different times of the year, the earth in its orbit around the sun moves at a high velocity toward Jupiter. At other times of the year, the earth moves away from Jupiter at a high rate of speed. Rømer would build on Cassini's observations during his studies of Jupiter, that the times between Jupiter's moon Io's eclipses got shorter as Earth approached Jupiter, and longer as Earth moved farther away. In 1676, after studying about 140 eclipses of Io, Rømer calculated that the speed of light must be 299,792 km/sec or 186,290 miles/sec (Born, pg. 91). We will examine this later.

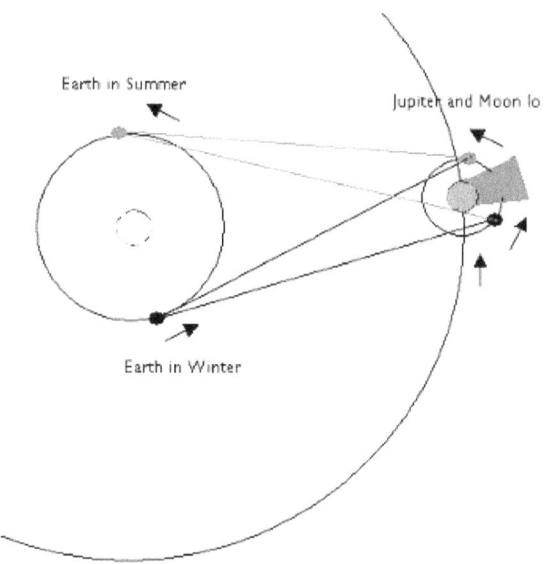

Jupiter's Moon Io

Star Dust

Now that the speed of light was known and considered finite, scientists began to question how light, magnetism, and gravity could traverse empty space. They wondered what light traveled through that limited its velocity. Empty space would have to be filled with a very fine imponderable substance they called the aether that would be the carrier or the medium in which light would travel, manifesting itself as the fields in which energy moved through empty space (Born, pg. 86).

Two theories of light propagation existed during this era, the **corpuscular theory** or **particle theory** and the **undulation theory**, otherwise known as **wave theory**. French philosopher **René Descartes** (1596-1650) revived the idea of old that the aether was the carrier of light as well as developing the law of refraction, also known as Descartes' law. **Sir Isaac Newton** (1643-1726) however was regarded as the author of the corpuscular theory, suggesting that light was something that moved out and away from bodies like "hot ejected particles" (Born, pg. 88).

Descartes' Luminiferous Aether would be an **elastic medium** made up of the smallest of particles, the God Particle or stardust if you will, permeating everything. A substance that oscillates, setting in motion other particles in waves, moving its energy through "empty" space. Diffraction, refraction and reflection of light all seemed to support this hypothesis that light moved in waves. This aether however **could not impede the motions or cause resistance to the planets or stars and yet must be rigid enough to transmit light** (Barbour, pg. 128). The earth therefore would move through this aether, possibly causing an **aether wind** to be detected across the earth's surface on its journey around the sun.

With Descartes undulation or wave theory, this aether would have to be considered an **elastic solid** due to the fact that light is transmitted as a transverse wave. Transverse waves cannot propagate through the fluidic motions of liquids and gases such as ocean waves and sound wave through air; this is because there is no resistance to lateral displacements of particles in these waves. Solid bodies propagate waves longitudinally and transversely; fore and aft, left and right and up and down as evident with the polarization of light (Isaacson, pg. 115).

According to the wave theory, light would theoretically travel through the aether in waves, very much like sound through air. A wave is defined as a traveling disturbance of coordinated vibrations that transmit energy through a substance **without net movement of that substance.**

There are two distinct types of waves; **longitudinal waves** and **transverse waves**. Both waves can travel in elastic solid mediums, however, waves traveling through fluids such as liquids and gases are always longitudinal waves. Longitudinal waves are waves in which oscillations are along the direction of wave travel such as in sound waves where there is a back and forth motion of these particles. The absence of rigid bonds between particles in a fluid prevents traverse waves from traveling through them. A transverse wave is a wave in which oscillations are perpendicular to the direction of wave travel such as a wave on a rope tied to a doorknob, electromagnetic waves and seismic waves traveling through the ground, calling for a somewhat rigid medium in order to conduct their energy (Tippler, pg. 465).

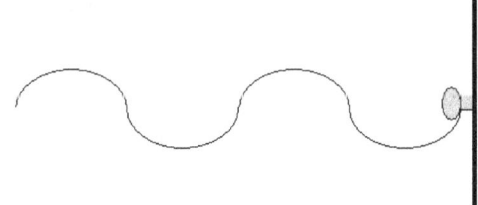

Transverse Wave

One way to view a longitudinal wave is to string single particles along in a line, somehow tied together with spring like forces. As one particle is forced away from another, an effect is felt on the particles to its left and to its right. One is pulled along for the short distance it moves and the other is pushed.

A chain event takes place as one particle pulls on another and one pushes on the other down the line in one direction longitudinally, also known as a compression wave.

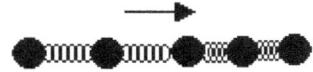

Longitudinal Wave

Now, visualize if you will, not only fore and aft but up and down and left and right motions in which all particles effect each other in all directions

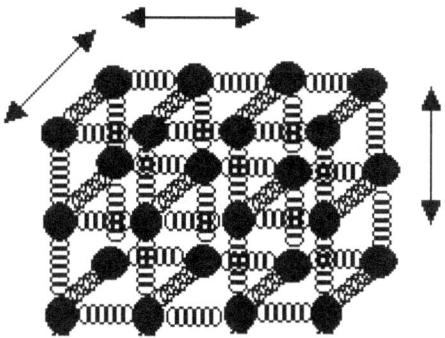

Transverse Wave

Polarization is the orientation of these transverse waves in light waves. Transverse light waves not only are propagated in a certain direction at the speed of light, they also oscillate perpendicularly to the direction of propagation, vertically and horizontally at the speed of light (Bord, pg. 326). The fact that light waves show properties of polarization reveals that light waves are indeed transverse waves. Therefore if light is propagated through a medium, and transverse waves can only be transmitted through solids, then **this medium that light moves through must be an elastic solid** like an "elastic jelly" (Psillos, pg. 131).

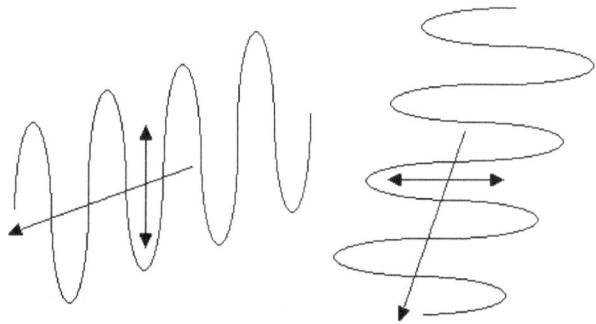

Horizontally and Vertically Polarized Transverse Waves

Light however seemed to also show signs that it was also a particle. Newton believed that through his studies with prisms, that light was made up of particles that moved at different velocities. Blue light particles for example, moved faster than red light particles, accounting for the different angles or refraction when passed through a prism. The fact that light was a particle also was manifested to him in that light traveled in a straight line. But Newton also recognized the phenomenon of diffraction and interference that could only be best explained when light was considered a wave, leading to a confusing view of **duality,** light being both waves and particles during the late 17th century (Bord, pg. 364).

Newton's particle theory however did not stand very long. **<u>Thomas Young</u>** (1773-1829), a brilliant Englishman, showed through the "**Double Slit**" or "**Two-Slit**" experiment, that light generated **interference patterns** and that light waves of different colors had different wavelengths. Through studies with polarized light, he discovered that light was a transverse wave, which supported the **undulation** or **wave theory** (Brown, pg. 87).

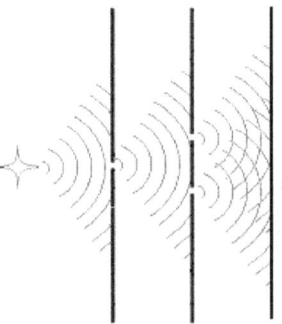

Double Slit Experiment

Star Aberration

During the eighteenth century, the Copernican heliocentric view of the universe had generally been accepted although there was very little empirical data to support it (Born, pg. 10). It was for this that Galileo in 1633 had risked martyrdom for the sake of what he believed to be truth.

In December 1725, one year before Newton's death, English astronomer **<u>James Bradley</u>** (1693-1762) began a prolonged observation of Gamma Draconis (also known as Eltanin), a star in the Draco constellation. This constellation is a group of circumpolar stars that remain in the night sky all year around and which passes directly overhead nightly where he lived in London. The study was an effort to the view the parallax of this star, using triangulation to determine its distance from the earth.

Throughout his observation viewing **Gamma Draconis** through a telescope fixed to the chimney on the roof, Bradley found that he had to continuously adjust his telescope to a different position almost nightly. Over a six months period, the position of the star seemed to move one direction then the other from its central position with a maximum variation of approximately 20.5 arc seconds. Unlike the circular motion expected due to the earth's tilt on its axis, the star seemed to move in an elliptical pattern. Bradley discovered that this angle of "aberration" depended on the ratio between the orbital velocity of the earth and the speed of light. Knowing the velocity of the earth, 30 km/sec, Bradley was able to determine the velocity of light very close to the calculations of Ole Rømer back in 1676 (Van der Kamp, pg. 27). This interpretation seemed to work well with either the corpuscular theory or the wave theory.

Let's look at this in another way. A tall box is designed that is totally enclosed except for a pinhole in the top that allows sunlight to shine on a small point on the floor. The box being on earth is moving at 30 km/second in its orbit around the sun. As the light shines through to the bottom of the box, the box moves, causing the light to strike off-center of the bottom of the box.

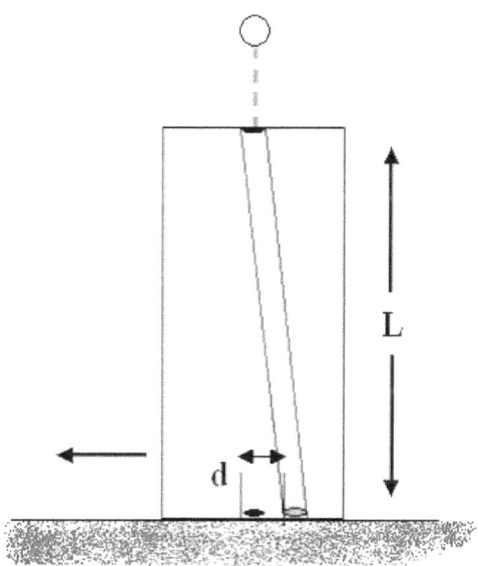

Tall Box Analogy

With the height of the box = L and the distance off center = d, we can use the formula $d = v \times l / c$. The speed at which the earth goes around the sun is: $v = 30$ km/s, therefore the speed of light will be $c = 299,792,458$ m/s (Born, pg. 95). This ends up being a very small numerical value; however, through the multiplication effects of a telescope's lenses and through fine adjustments, this amounts to 20.5 arc seconds in displacement of the telescopes angle in the direction of the earth's motion.

$$\frac{d}{l} = \frac{v}{c}$$

Star Aberration

Star aberration or **stellar aberration** is caused by a slight displacement of a stars visible position when viewed through a telescope; this change in position is independent of the stars distance, unlike stellar parallax. One way to visualize this is by considering what happens to rain as it falls down a vertical stovepipe carried by someone walking in the rain. If the person is stationary, the rain falls straight through to the bottom of the pipe. If the person carrying the pipe begins to walk, the rain falling through the stovepipe will begin to hit the inside wall of the pipe. The pipe would have to be held at an angle in order for the rain to fall straight through. Furthermore, the faster the person walks, the more tilted the pipe must be held to allow the rain to continue falling through. Another way to view this would be a person holding an umbrella and walking. The faster one walks; the more tilt is needed on the umbrella. This stellar aberration also can be observed only with the movement of the observer and completely disappears if the observer is stationary and the light source is moving. A way to understand this is for example; a cloud that is moving overhead does not cause the rain to fall at an angle when there is no wind – the stove pipe would still have to be held vertically.

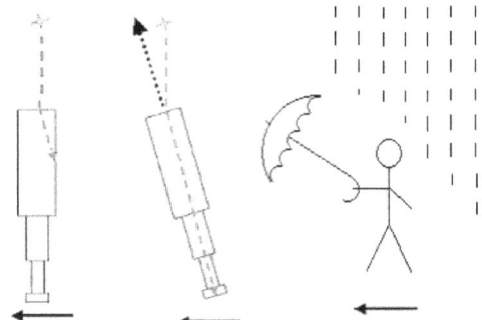

Stellar Aberration Displacement

A star viewed on the earth's ecliptic plane will vary in its position in a straight line. As an observer tilts the telescope north towards circumpolar stars, these changes in apparent position of the stars over 6 months turn into ellipses and then circles directly over the poles. The sun has a truly special case as its apparent position, staying at a constant 20.49" arc seconds. This was the empirical proof needed that the earth was in motion around the sun and that the Copernican heliocentric view of the universe was correct.

Stars on the earth's ecliptic plane can only be viewed for part of the year, becoming visible at dusk, rising at an angle in the night sky due to the earth's tilt and fade from view at dawn. These stars will change in position from night to night, first noticeable at sunrise in the east, moving towards the west a little every night until they are no longer visible being below the western horizon; these stars are then in the daylight sky and cannot be viewed. The **apparent position** of these stars when they first are visible does not shift, then as the months go by and the star rises in **Right Ascension** (RA) or **(α)**, east to west, its apparent position moves in a straight line until it is most noticeable when the earth's rotation is 90 degrees to the stars light.

As the star moves into **Declination** (DEC) or **(δ)** in the west, this apparent position is more noticeable and the star seems to accelerate back to its true position.

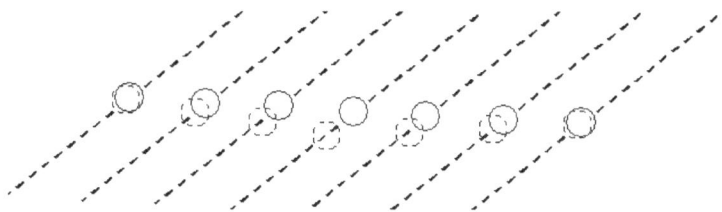

Apparent Position (ecliptic plane)

Stars in the Polar Regions are called circumpolar stars and can be viewed all year in their hemispheres. Northern circumpolar stars rotate around the North Star, Polaris, and can be seen anywhere between the equator and the arctic circle all year. Below the equator, Polaris is not visible, dropping below the northern horizon; above the Arctic Circle, these stars can only be viewed during the winter, as the sun does not set in the summer. These stars should rotate in a circular pattern during the year due to the tilt of the earth's axis but move in an elliptical motion due to stellar aberration; Polaris is the only star that remains in a circular pattern. The further south the star is, the more elliptical its motion becomes. These circumpolar stars shift in their apparent position in one direction in the winter months and in the other direction during the summer months, depending on their position in the night sky. This apparent shift in position is most noticeable when the earth's orbit is at 90 degrees to the starlight.

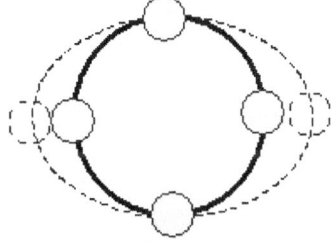

Apparent Position (circumpolar stars)

The star being viewed is so far off that the angle of starlight reaching the earth is the same throughout the year. Most stars can only be viewed for a part of the year due to their position. Stars lower on the horizon move into the daylight side of the earth for six months of the year or so and cannot be viewed; the sun will even eclipses some of these stars. When these stars first appear, they become visible just before dawn (position A) and starlight hits the earth parallel to its movement. As they continue orbiting the sun, the angle at which light from the stars hits the earth is at 90 degrees to the earth's orbit (position C). After six months or so the star fades back into daylight becoming last visible at sunset where the starlight hitting the earth is again parallel to its movement.

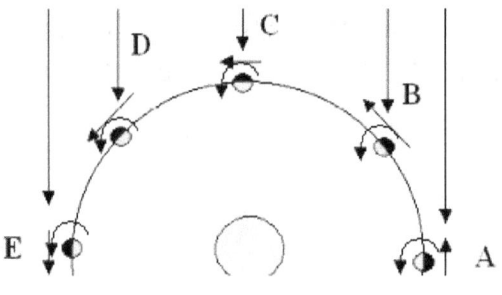

Earth's Orbit

François Arago (1786-1853) a French mathematician and physicist, assumed along with other scientists of that day, that the speed of light from a large star would be slower than light from a small star due to its massive gravitational pull on the particles of light emitted from it. In 1810, Arago, in appreciation of Newton's **refraction theory**, came up with a simple idea to detect various light velocities from stars. Since the refraction angle of light will be different for particles of light moving at different speeds, a prism mounted on a telescope when focused on different stars should be able to differentiate slow light from fast light. Slow light from large stars would refract at sharper angles than fast light from smaller stars. Stars moving towards the earth and stars moving away from earth should have different speeds of light coming from them and should have different **refraction indexes**. This experiment failed to show any deviations, indicating that light coming from all stars was exactly the same speed through the prism – even the various colors.

Arago States:

"By examining the receding tables attentively, one finds that the rays of all stars are prone to the same deviations...

This result seems to be, with the first aspect, in manifest contradiction with the Newtonian theory of the refraction, since a real inequality in the speed of the rays however does not cause any inequality in the deviations, which they test."

(Arago, pg. 562-563)

Augustin Fresnel, (1788-1827) French physicist, in September of 1818 responds to Arago's discovery in a letter:

"By your fine experiments on the light from stars, you have demonstrated that the movement of the terrestrial globe has no noticeable influence on the refraction of rays emanating from these stars..."

"You have inspired me to investigate whether the result of these observations could be more easily reconciled with the system which takes light as consisting of vibrations in a universal fluid. It is all the more necessary to explain these results using this theory, since it must apply equally to terrestrial objects; for the speed with which the waves are propagated is independent of the movement of the bodies from which they emanate."

(Fresnel, pg. 57–66)

Throughout the 1800's more studies were performed in an effort to find the aether and prove that the speed of light was finite. In **1849 - Armand Hippolyte Louis Fizeau** (1819-1896) (who had predicted red shift) almost 200 years after Rømer in 1848 used a series of mirrors and half-silvered mirrors over a distance of 8.6 Km (5.36 miles) to determine light speed. Using a **spinning toothed wheel** as a timing device through an aperture, Fizeau calculated the speed of light at 313,300 Km/sec or 194,684 miles/sec (Born, pg. 96). Other experiments such as those with the *interferometer* and experiments involving light wave interference phenomena confirmed this.

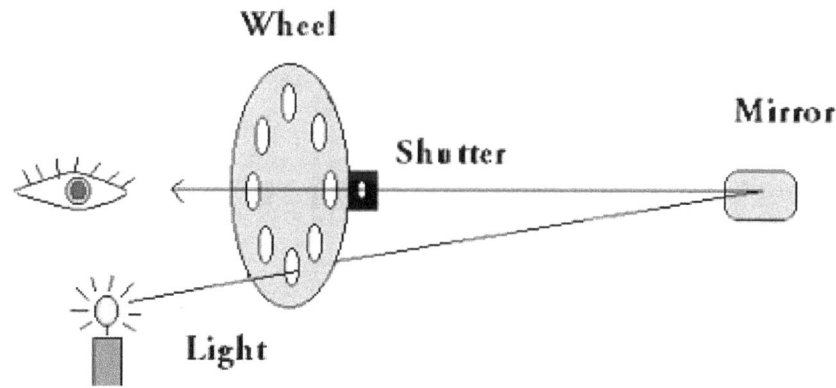

Fizeau's Toothed Wheel

1850 - Jean Bernard Léon Foucault, (1819-1868) improved upon this method bouncing light off a spinning mirror to a stationary mirror. When the light returned to the rotating mirror it reflected at a slightly different angle, which was used to determine the lights velocity improving accuracy by about 1,000 miles per second. Using the interferometer, **Albert Michelson** refined this experiment determining that light speed was 299,796 Km/sec or 186,293 miles/sec, within 3 miles per second of Ole Rømer's measurement in 1676 almost 200 year earlier (Born, pg. 103).

Foulcault's studies had shown that the velocity of light in water was much slower than that of light moving through the air. Bradley had suggested that since light traveled slower in water when compared to air that a **telescope filled with water** should increase the stellar aberration. In 1871, more than one hundred years later, **Sir George Biddell Airy** (1801-1892) demonstrated that stellar aberration occurs even when a telescope is filled with water and that there was no change in the aberration angle. This null result suggested a lack of universal ether (Van der Kamp, Walter pg. 30).

In **1887** - **Albert Michelson** (1852-1931) and **Edward Morley** (1838-1923) set out to prove the existence of the aether using an **interferometer.** It was assumed by the experts of the day that the aether was a **rigid substance**, which was stationary in space and was the *medium* in which light traveled. It was believed that as the stars and planets traveled through space, they passed through this substance creating an "**aether wind**" that should be detectable with some sort of device. Michelson and Morley set up an interferometer (a devise that utilized mirrors and detectors to measure the speed of light) in such a way that as the earth passed through the aether wind in space, distortion or interference fringes would develop on the detector. The theory was that the speed of light would be different in each of the arms of the detector, one pointing into the aether wind (the direction the earth was moving in space) and the other perpendicular to the assumed aether wind. **No wind was detected** (Ditchburn, pg. 315). We will also examine this a little later.

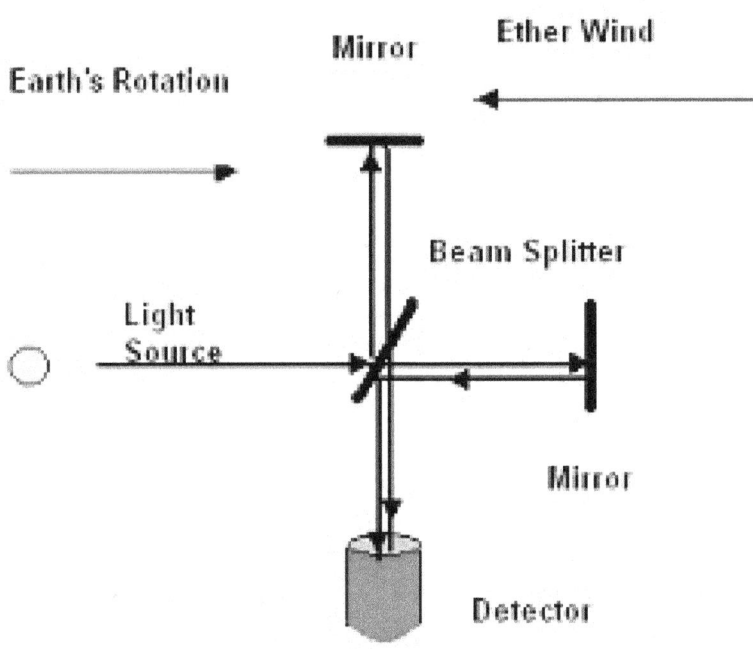

Michelson's Interferometer

The famed Michelson – Morley Experiment in 1887 with the *interferometer*, and other experiments involving light wave interference phenomena confirmed to scientists of that day that there was no aether. Not only did they seem to show that there was no aether, they had also discovered what they believed to be the "Law of the Constancy of the Speed of Light." This constant is expressed as "c" and is measured at 299,853 km/sec. or 186,328 miles/sec. This measurement of the speed of light in air would not vary and is the same velocity no matter what the observer's velocity is or what velocity of the emitter is, either moving towards or away from each other (Wolf, pg. 52).

Partial Aether Drag Theory

*"If one supposed that our globe impresses its motion on the aether in which it is enveloped, one would easily conceive why the same prism always refracts light in the same manner, whatever the side from which the light may arrive. **But it seems impossible to explain the aberration of stars by this hypothesis**: I have not been able, so far at least, to account for this phenomenon clearly except by supposing that the ether passes freely through the globe, and that the speed communicated to this subtle fluid is only a small fraction of that of the earth, not exceeding one hundredth of it, for example"*

(Fresnel, pg. 57–66)

Fresnel now was suggesting that the aether or medium through which light traveled, must have been sticking to the earth, being dragged along. As light from the stars would enter this medium at different velocities, it would enter the earth's atmosphere at one speed. It seems however, that if the aether drag hypothesis were true, then stellar aberration could not occur because the light would be traveling in the aether, which would be moving along with the telescope (Van der Kamp, pg. 30). Fresnel at first saw this as contradictory but then suggested that material with low refractive indexes, such as air, must produce almost no resistance at all to stellar light allowing stellar aberration to occur. However, substances with a higher refractive index, such as water and glass, produce more resistance, thus allowing for Arago's experiment that indicated the constancy of the speed of light.

Light would therefore travel unimpeded through the air and be uninfluenced by the movement of the earth. Once light enters into glass or some other medium of higher refractive indexes, it would therefore be constrained or locked into the new medium and travel at a different (slower) velocity within that medium. The Fresnel **partial drag theory** was widely received during the last part of the nineteenth century. This both allowed for the constancy of the speed of light and for star aberration.

The aether drag theory suggested that the velocity of light would be affected by a moving medium. Fresnel projected that the aether's density in a substance was different than that of the aether in empty space and depended on the refractive index (*n*) of the substance it was traveling through. Air would have very little resistance to the aether but substances like water and glass would have a lot of resistance. Here c = the speed of light, n = refractive index, and v = to the velocity of the medium (Ditchburn, pg. 334).

$$c' = c_1 + v\left(1 - \frac{1}{n^2}\right)$$

Fresnel's Convection (Aether Drag) Coefficient

In an effort to find some variation in the stellar aberration, Sir George Biddell Airy in 1871, as noted previously, carried this idea a little further with the use of a water-filled telescope as suggested by Fresnel in his 1818 letter to Arago. Theoretically as light would travel through the water as compared with air, refraction of light in the water would affect stellar aberration in some degree. **It did not.**

The Dilemma

The speed of light was now known. However, experiments such as the famous Michelson – Morley Experiment indicated that the velocity of light was always the same, independent of the motion of the emitter or the observer. The "simple emission theory" suggests that light given off by an object should move at a velocity relative to the movement of the emitting object and/or the movement of the observer, however this was not the case. The aether drag theory tried to answer this with an aether that was being dragged along with the earth acting as a medium, causing all light from heavenly bodies to be perceived at one velocity – like sound is through air. Stellar aberration however contradicted this by proving that light was not being carried by any medium, at least not in the earth's atmosphere. Fresnel partial drag theory tried to correct this by stating that aether drag was dependent on refractive indexes and that air had little effect on light; denser object such as glass and water however would carry light at varying degrees. Airy's water filled telescope experiment in 1871 seemed to contradict this.

In 1913 while studying **Binary Stars** or **twin stars** that orbit around each other, **Willem de Sitter** (1872-1934) a Dutch mathematician, physicist and astronomer, suggested that stars in these systems should have an orbit that emitted light at different speeds (c + v or c − v). The problem was that the "faster" light given off during the orbit of a star towards the earth would be able to catch up with and even overtake the "slower" light emitted from the other star in the system retreating away from the earth. Theoretically, this difference in light velocities would cause the stars light to be scrambled and out of sequence, distorting the light seen from these binary stars (Einstein, pg. 18).

The fact that there are no distortion means that we can be positively sure that the speed of light coming from both stars is the same from the instant it left the stars. With the assumption that the movement of observer is equivalent to the movement of the stars, this leads us to the conclusion that the speed of light does not depend on the speed of observer or the movement of the stars; the speed of light will always be 300,000 km/sec to the observer detecting the light (Sokolov, pg. 2).

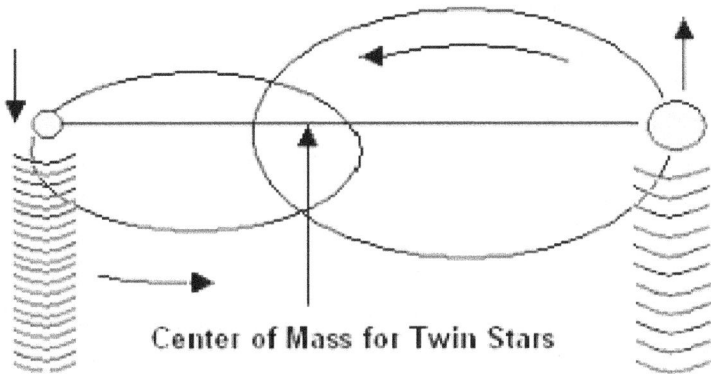

Center of Mass for Twin Stars

Doppler Effect Twin Stars

Although the light coming from these stars travels at the same velocity, the wavelength of that light is not the same. When one of the twin stars is moving towards the earth, the wavelengths compress and become shorter, towards the blue in the light spectrum, known as Blue Shift (similar to the effect of a whistle emitted from the caboose to an observer on the train). When one of the twin stars is moving away from the earth, the wavelengths are shifted to longer wavelengths into the red portion of the light spectrum, known as Red Shift (similar to the effect of a whistle emitted from the engine to an observer on a train) producing the "Doppler Effect" (Bless, pg. 235).

This now creates a dilemma and raises some important questions. If light is made up of particles called photons or "hot ejected particles" as Newton would put it, and was leaving a star moving towards a detector or an observer at the speed of light, what is it that slows the photons down? If the star emitting the light being detected is moving away from the detector, what speeds up these photons? The photons must be moving at the speed of light in reference to the moving star that they came from yet be moving at the speed of light in respect to the detector or observer. How can this be?

Before the 1900's it was natural for almost everyone to believe that light must be transmitted through the material called the aether, which permeated everything. This aether would fill all of the space between the stars and the earth; otherwise light could not be visible coming from these stars. If light moved in waves, as it seemed to, just as there must be air for sound to move through, there must also be a medium for light to move through. In his book **Opticks**, published in 1704, Sir Isaac Newton wrote:

"Is not the heat of the warm room conveyed through the vacuum by the vibrations of a much subtler medium than air, which after the air was drawn out remained in the vacuum? And is not this medium the same with that medium by which light is refracted and reflected, and by whose vibrations light communicates heat to bodies, and is put into fits of easy reflection and easy transmission?"

(Newton, Sir Isaac, Book III, Part I, pg.323)

We know that sound has a fixed velocity in air at a given temperature: would this also be the case with light traveling through this aether? If this aether was present throughout space, as the Earth rotates on its axis and moves in its orbit around the sun, should not this aether be possibly detected as an eerie **wind** blowing through the earth, affecting the wavelengths of light moving in specific directions. Light waves moving with the earth's rotation it seems should be compressed and light waves moving perpendicular to the earth's rotation should not be affected.

As previously discussed, the theory behind the interferometer built by Michelson and Morley was that, as the earth rotated and moved in its orbit around the sun, the aether wind that was stationary in space would flow through the detector at a specific angle. A light beam would be split with a half-silvered glass plate and be reflected off two mirrors. One beam would be pointed into the aether wind (presumed to be flowing in the opposite direction of the earth's rotation and movement through space) and the other reflected perpendicular to the aether wind. The light beamed into the aether wind would slow down compressing the waves, making them shorter. When these waves enter the detector they would be out of phase with the perpendicular beam that was not affected causing distortion on the detector. If there was no distortion there evidently was no aether wind.

The detector was build and floated on pool of liquid mercury to dampen out any noise and vibrations caused by the surroundings. After several experiments with no distortion in the detector, it was determined that there was no **aether wind** (Collins, pg. 37). The Michelson Morley experiment seemed to confirm that light takes the same time to travel each path irrespective of the motion of the observer or detector.

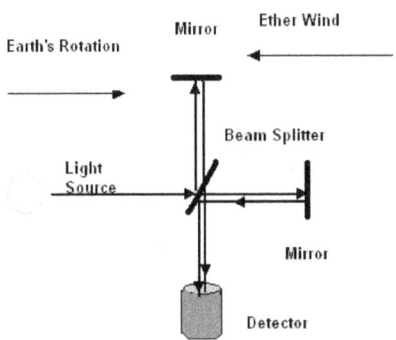

The Michelson Morley Experiment

In 1895, Dutch physicist **Hedrick Lorentz** (1853–1928) theorized that the "null" result obtained by Michelson and Morley must have been caused by an "effect of contraction" made by the aether on their apparatus and introduced the length contraction equation (Collins, pg. 38). Where L is relativistic length and L_0 is length of object at rest. The idea here is that the aether must still exist but that high-speed movement through space distorts space enough to nullify any measurements that could detect the aether; the higher the velocity the more contraction of space. This is now known as the "Lorentz Contraction".

These discoveries at the turn of the century changed science; from this, Albert Einstein developed his "Theory of Relativity".

$$l' = l \sqrt{1 - \frac{v^2}{c^2}}$$

Lorentz Contraction Equation

Albert Einstein (1879-1955) was born in 1879 into a nonreligious Jewish family in the city of Ulm, southern Germany (Isaacson, pg. 11). He was born the same year of Scotsman James Clerk Maxwell's death, one of the leading scientists of the day in the field of Electricity and Magnetism, an area in which Einstein would become well versed. Apparently stubborn, young Einstein did not speak until the age of 3, waiting till he could talk in full sentences. He loved playing with puzzles, working geometry problems and enjoyed toying with magnets. Later on as a teenager, due to conflict with his teacher, Einstein would end up dropping out of high school, but was still able to enroll in a Swiss university. Although being considered "lazy" by some of the professors there, Einstein soon became well-schooled in the area of physics; a field he would soon pioneer and later would develop the famous "Theory of Relativity" (Bartusiak, pg. 29-30).

The Special Theory of Relativity

Albert Einstein began to wrestle with these seemingly contradictory findings, struggling with the properties of light whether it was wave or particle and the constancy of the speed of light. After years of study, he came to a conclusion that if the velocity of light was constant, no matter what the velocity of the observer towards or away from a light source, or the velocity of the object emitting the light (as indicated by the famous Michelson – Morley Experiment in 1887) that this velocity must both change time and space. In order to fully understand Einstein's "Theory of Relativity", one needs to understand "relative velocity". To measure the velocity of any object, you must have a means of measuring distance and a means of measuring time. Einstein visualized this in his writings with a measuring rod (yard stick, tape, whatever) and a clock (Kafatos, pg. 24). Today, we can use a baseball and a radar speed detector as an example.

If the baseball is thrown towards the stationary radar detector at 100 mph, the detector will detected the baseball at that same velocity - 100 mph. Now, let's put the radar detector in a vehicle and drive at 50 mph towards the baseball being thrown at 100 mph. This moving baseball can now be viewed as having two speeds, one speed relative to the pitcher who threw the ball at 100 mph and the speed of the baseball relative to the vehicle being detected at 150 mph; the combination of the vehicles speed of 50 mph added to the speed of the ball relative to the pitcher of 100 mph.

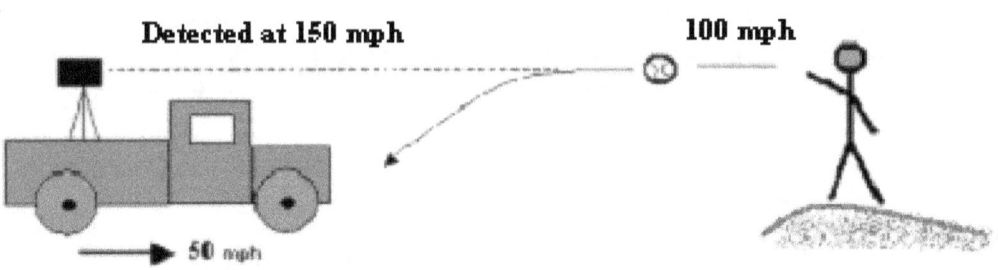

Baseball at 150 mph

On the other hand, if the detector on the vehicle is moving away from the baseball at 50 mph and the baseball itself is thrown at 100 mph relative to the pitcher, the baseball will be detected by the radar detector at 50 mph (as viewed from the moving vehicle) (Bord, pg. 448).

The velocity of the baseball has to be measured relative to something, in this case, a radar speed detector moving in a vehicle or the velocity of the baseball relative to the pitcher. During the pitch, the ball is traveling at a different velocity relative to the pitcher who threw the ball and relative to the moving radar detector in the vehicle; it's all a matter of perspective. A bystander or passerby will have another relative velocity depending on their direction of movement and velocity relative to the baseball.

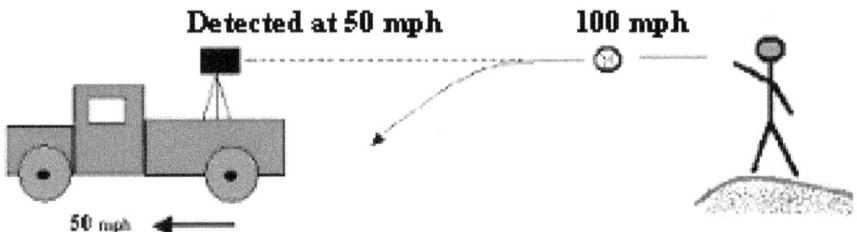

Baseball at 50 mph

The problem now is that when measuring the velocity of light, no matter what the velocity of the observer (radar gun in this case) towards or away from the light source (baseball), the measurement of the velocity of the light from this object remains the same (Born, pg.231-232).

We can use the earth as an example of this in much the same way as the baseball. As the earth orbits the sun, its mean orbital velocity through space is approximately 30 kilometers/sec. (18 miles/sec) or 64,800 miles per hour. As the earth in its orbit moves directly toward a star, (assuming that the star is stationary in space) the velocity of the light detected from that star, *should* be the sum of the velocity of the earth toward the star and the speed of the light from that star, similar to the baseball and radar detector moving towards each other (simple emission theory). On the other hand, as the earth moves away from the star in its orbit around the sun, the speed of the light detected *should* be the difference between the velocity of the earth in its orbit away from the star and the speed of the light from the star, similar to the baseball and radar detector moving away from each other (Born, pg. 132).

Movement of the earth towards a star should be:

EXAMPLE 1: 186,000 m/sec +18 m/sec = 186,018 m/sec

Movement of the earth away from a star should be:

EXAMPLE 2: 186,000 m/sec - 18 m/sec = 185,982 m/sec

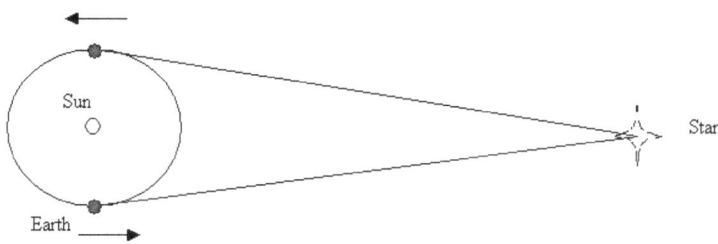

Earth and Star

But this is not the case. Sir Isaac Newton's **Classical Mechanics** and this new Theory of Relativity are now in conflict. Einstein himself said:

"Who would imagine that this simple law (the constancy of the velocity of light in a vacuum) has plunged the conscientiously thoughtful physicist into the greatest of intellectual difficulties?"

(Einstein, pg. 19)

"Law of the Constancy of the Speed of Light" was now stated:

"The velocity of the propagation of light cannot depend on the velocity of motion of the body emitting the light" (nor the motion of the observer detecting the light).

(Einstein, pg. 18)

No matter what the velocity of a detector towards or away from the star, or the velocity of the star itself towards or away from the earth (or detector), the speed of light coming from the star into the detector is always the same, 186,000 m/sec. The math doesn't seem to work – according to Einstein, space and time must be being warped.

EXAMPLE 1: 186,000 m/sec +18 m/sec = 186,000 m/sec

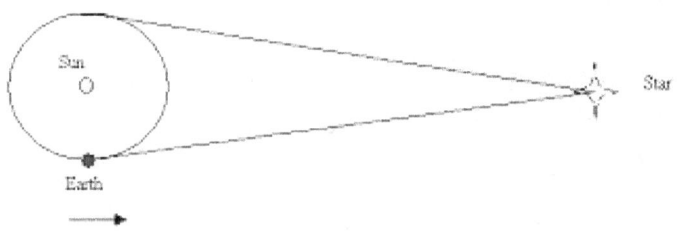

Earth Traveling Towards a Star

EXAMPLE 2: 186,000 m/sec - 18 m/sec = 186,000 m/sec

Earth Traveling Away From a Star

To add to this perspective, in addition to the earth's rotation around the sun being 30 km/sec (18 mi/sec) in reference to a star in our galaxy, the movement of the solar system rotating around the center of the Milky Way galaxy is about 250 km/sec (140 mi/s). The Milky Way itself is traveling approximately 300 km/s (185 mi/s.) in reference to other galaxies in the "Local Group". Most galaxies are spreading out away from each other at enormous velocities in an ever-expanding universe. The great **Andromeda Spiral Galaxy** however, one of the Milky Way's nearest neighbors, is actually moving toward the **Milky Way** and earth at about 50 km/s (about 30 mi/s.)

The General Theory of Relativity

Albert Einstein would then build on "Special Relativity" with the knowledge he gained in studying gravitational fields. He would go on to explain the universe and everything within it as four-dimensional "Time-Space". In three-dimensional space, the x axis would be the fore and aft, the y axis would be the side to side and the z would be the up and down with respect to the "Fixed System" reference; the fourth dimension is "time", creating Einstein's universe of Space-Time. Velocity could then be explained as the change in position over time. From this perspective, Einstein would develop the theory of the constancy of the speed of light, the theory of gravity as well as explain gravitational redshift and even time dilation.

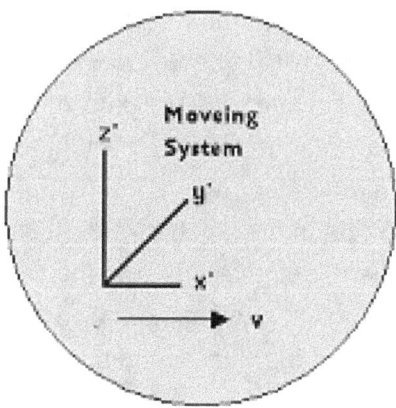

Time- Space

Gravity as Einstein would see it, exists due to "time-space curvature (Isaacson, pg. 193); as if the fabric of space was being warped. Picture if you will four people holding a sheet at the four corners tightly, then someone sets down a bowling ball in the center of the sheet causing the center of the sheet to sink. A marble is then rolled onto the blanket; as it rolls, it spins around the bowling ball a few times in a tight spiral until it stops, down between the bowling ball and the blanket. From this premise, Einstein would develop his famous "**Field Equation of Gravity**":

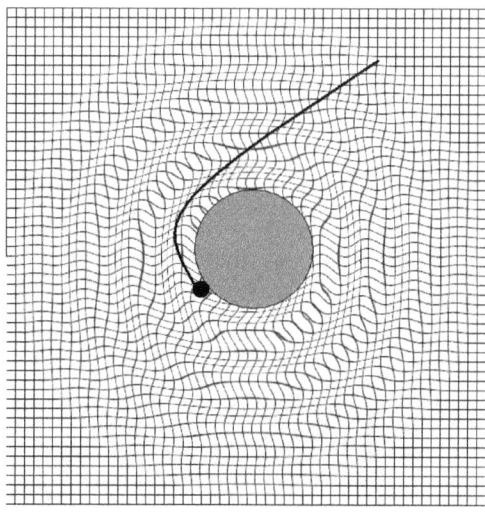

Gravity

Einstein's "Field Equation of Gravity":

$$R_{uv} - \frac{1}{2} g_{uv} R = T_{uv}$$

The left side of the field equation of gravity represents a geometric gravity field of space-time and the right, the distribution of mass-energy. Both sides create a balance, the more the mass-energy, the greater the warping of space-time; one affects the other and there cannot be the one without the other. Objects such as planets simply naturally follow these curves in time-space, as well does light. Einstein's new formula actually doubled his original formula in the amount of deflection that gravity would cause a ray of light to bend. Later we will see that this was the case; in 1919 when astronomer Arthur Eddington photographed a stars new apparent position during a solar eclipse, where the suns gravity actually changed the stars apparent position in the sky (Bartusiak, pg.45).

We will also see that not only is motion influenced by this curvature of space, but that time also is affected (Einstein, pg. 81). Recent technologies such as the Global Positioning System (GPS) must adjust for time dilation due to the earth's gravitational field and the satellites velocity (Bartusiak, pg. 55).

Relativity Re-examined

The principle of relativity is not really new. Galileo described possibly the first principle of relativity in his book, *Dialogue Concerning the Two Chief World Systems*. Here he performs a mind or thought experiment with various elements in order to describe movement of objects relative to other objects. Galileo illustrated this with a ship floating on a smooth sea with an observer, butterflies, a bottle dripping water and fish swimming in a bowl in a closed cabin. The observer will first observe these randomly moving objects while the ship is floating stationary and then while the ship is sailing. Here, there is no noticeable difference until the observer is allowed to look outside and see the direction of the ships motion. The movement of the fish, butterflies and even the bottle dripping water, all move relative to each other with no change even when the ship as a whole is moving or not, relative to land and sea, showing that all uniform motion is relative (Galileo, pg. 187). This is the premise of Albert Einstein's Special Theory of Relativity and of General Relativity.

Another way to look at "relativity" is to look at how sound travels through air and the "Doppler" affect. For instance, let's say there is a train traveling at a constant rate with one observer on the train midpoint between the engine and the caboose. On the embankment next to the passing train is another observer, standing still, also midpoint between the engine and the caboose.

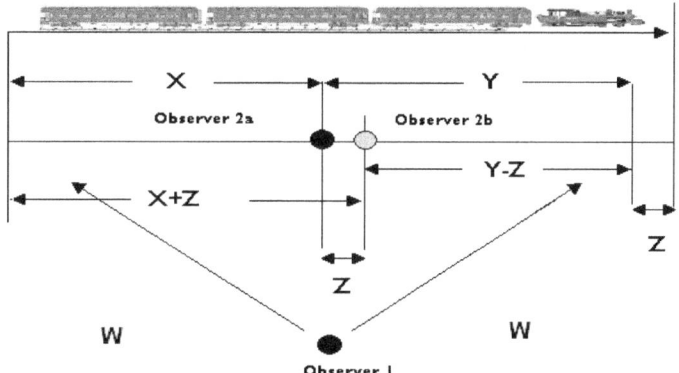

Einstein's Train - Sound

The engine sounds its whistle simultaneously with the caboose whistle (both engineer and conductor have their clocks perfectly synchronized). <u>Observer 1</u> on the embankment next to the passing train (see Figure above) is equidistant from the engine and caboose as sound travels through the still air to the stationary observer (distance W); he therefore hears both whistles <u>at the same time</u>. Observer 2, on the train however, will have a different perspective. Observer 2 will move distance Z during the time it takes the sound to first reach him from the engine and caboose. The sound traveling to him from the engine will cover distance Y-Z while the sound traveling to him from the caboose will travel X+Z distance.

The speed of sound varies depending mainly on the, temperature and the *medium* through which it travels. At a standard atmosphere and at a temperature of 20^0 C sound travels at 344 meters per second in dry air or for a standard day, 59^0 Fahrenheit (15^0 Celsius), 761 miles per hour (Bord, pg. 225). We shall therefore use 761 mph for the speed of sound in this mind experiment.

The train passing the embankment is traveling at 39 mph and is one mile long; therefore distances X and Y are a half-mile. That is 761 MPH = 12.7 miles per minute (MPM) or .211 miles per second (MPS). To convert miles per second to seconds per mile we inverse MPS into seconds per mile (SPM) by 1/ .211 which is equal to 4.731 seconds per mile or 2.3655 seconds over a half-mile (Y).

The question now is how far did the Observer 2 travel during the time it took sound to cover distance Y and X? This would be distance Z. By simple calculation we discover that Observer 2 traveled Z distance in 2.365 seconds while traveling at 39 miles per hour (57.2 ft. per second), therefore, 2.365 X 57.2 = Z = 135.278 ft. If half a mile (Y) = 2640 ft., then Y-Z = 2505 ft. and X+Z= 2775.278 ft. If sound travels at .211 miles per second, it then travels at 1116 ft. per second. Y-Z = 2.25 seconds and X+Z = 2.37 seconds a difference of .12 seconds.

Now, what if this was a high speed Bullet Train traveling at 139 mph (203.866 ft. per second)? Therefore, Z distance for Observer 2 = 2.365 X 203.866 = Z = 482.14 ft. If half mile (Y) = 2640 ft., then Y-Z = 2157.86 ft. = and X+Z = 3122.14 ft. If sound travels at .211 miles per second, it then travels at 1116 ft. per second. Y-Z = 1.93 seconds and X + Z = 2.80 seconds, a difference of .87 seconds. In other words, Observer 2 at midpoint on the train will hear the engine whistle almost one full second before he hears the cabooses whistle.

The "Doppler Effect" is another phenomenon associated with sound and movement. We have all heard the train whistles pitch change as the train approaches us and then passes us – a higher to lower frequency.

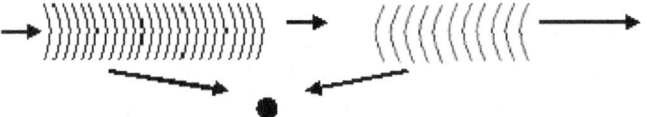

Doppler Effect on Observer 1

Although the Observer 1 on the embankment hears both whistles sound at the same time, he hears them at different pitches. As sound emits from the engines whistle, although its pitch is the same as the cabooses, its wavelength is lengthened or stretched as the train moves forward away from the observer, increasing the distance between wave peaks which in turn causes the pitch to be lower in tone. This is because sound transitions from a moving whistle into stationary air. Likewise the wavelength of the sound emitted from the caboose whistle moving towards Observer 1 on the embankment is compressed and shortened, increasing in pitch as the caboose moves forward into the still air.

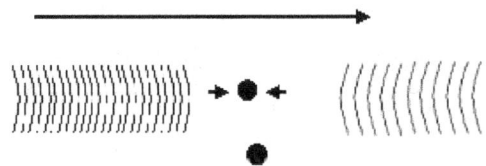

Doppler Effect on Observer 2

Observer 2 on the moving train however will hear the pitch from the engines whistle and the pitch of the caboose whistle at the same pitch, but not at the same time. Although the wavelength of the sound coming towards him has a longer wavelength, as he moves forward into the sound, the wave peaks reach his ears at a faster rate and he hears no difference. Sound also coming from the caboose, being compressed by the cabooses movement through still air will also be perceived as the same pitch as he moves away from the compressed sound waves, lengthening their perceived wavelength.

Einstein explains relativity similarly and quite eloquently with his "railway carriage" and "embankment" analogy in his book, Relativity, Special and General Theory (Einstein, pg. 10). In this comparison Einstein's train is also traveling at a constant rate along an embankment with observers on both the train and embankment. Lightning strikes simultaneously at the front and at the rear of the train while the observers on the embankment and on the train record their observations.

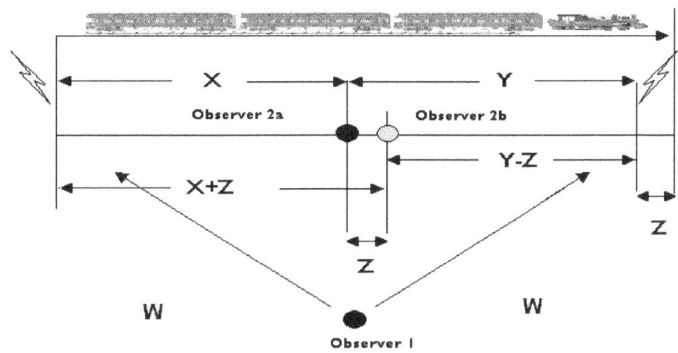

Einstein's Train - Light

The Observer 1 on the embankment records that the lighting struck simultaneously at the front and rear cars of the train, but Observer 2 riding in the center of the train records that the lighting at the front stuck first.

As the lightning struck, the train continued to move forward at a constant rate. By the time the light reached Observer 2 from the front of the train, the observer had moved distance Z forward to the Observer 2b position. The distance the light had to travel was then Y-Z. Likewise, Observer 2 moving Z distance away from the lightning striking at the rear car causes the light to travel distance X+Z to the Observer at the 2b position, therefore allowing the assumption that it struck later. This is exactly the same scenario and outcome as our thought experiment with the train whistle.

Rømer's Train

This same scenario (covered earlier with Einstein's Train) occurred with Dutch astronomer Ole Rømer in 1676 as he observed the movements of Io in orbit around Jupiter (Born, pg. 91). Observer "A" (Rømer) on the earth (the train) in the summer moved "Z" distance towards Jupiter while Io was in eclipse behind Jupiter; the light therefore had to travel Y-Z. Likewise in the winter (six months later) while the earth moved away from Jupiter, Observer B (Rømer again) moved Z distance away from Jupiter during Io's eclipse, causing the light reflected off of Io to reach Observer B over a distance of X+Z. Remember that Cassini, whom Roemer studied under, observed that Io's eclipses got shorter as Earth approached Jupiter, and longer as Earth moved farther away.

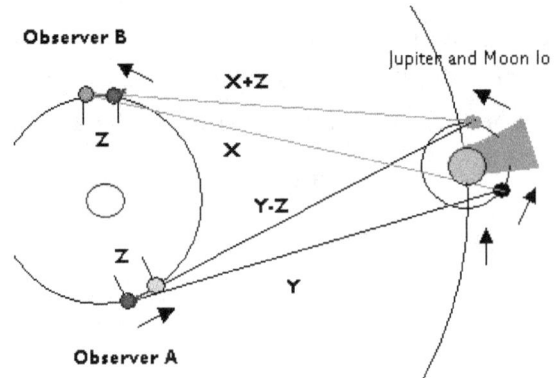

Jupiter and Io

Although this change shifted only seconds per day, to Romer this was a 17-minute change in Io's eclipse from summer to winter. This is due to the high rate of velocity of the earth towards Jupiter at one time of year (approximately 67,000 mph) while at the same rate of velocity away from Jupiter and moon Io months later.

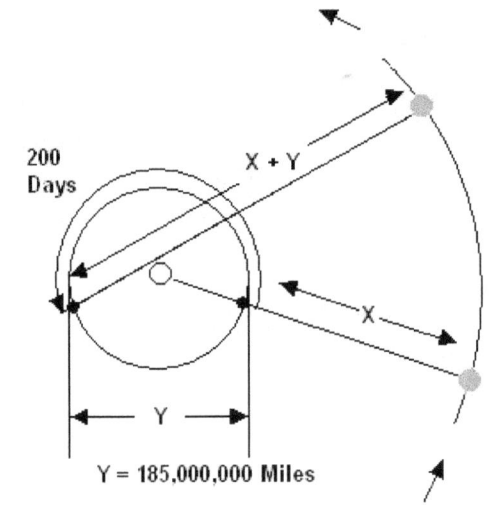

Rømer's Discovery

Prior to this point we have been focusing on the difference in velocity towards or away from an object (baseball, engine, caboose or moon). There is however an easier way to examine these changes. Keeping it simple, let's examine how Rømer calculated the speed of light. At point X where earth is closest to Jupiter, Rømer timed the orbit of Io around Jupiter. Two hundred days later when earth was at its furthest point away from Jupiter, Rømer took a new reading which was 1,000 seconds longer. Scientists of that day had determined that the earth's orbital diameter was approximately 299,793,000 kilometers through triangulation and calculations performed on other planets. Rømer determined that the additional time was due to the additional distance that the reflected light had to go – distance Y. If light took 1000 second longer to travel an additional 299,793,000 km, then through simple mathematics he calculated that the speed of light = 299,793,000/1000 seconds which is equal to 299,793 kilometers per second or 186,291 miles per second (Born, pg.93).

Light, Gravity and Relativity

The theory of light prior to Einstein assumed that light was not made of matter and that gravity would have no effect on light. Einstein's Theory of Relativity and the field equations he had created generated a new assumption, that light was a particle and that being created of matter, gravity must attract light, curving its path as it passes close to a large stellar mass such as a star. A stars apparent position should shift by about 1.7 arc seconds according to Einstein's calculations due to the suns gravity (Bartusiak, pg. 47).

Einstein set out to observe and record how much a stars light was deflected by the sun due to the changes in space-time as light passed through its strong gravitational field. This was not an easy task as stars are not normally visible due to the suns brightness, so in 1919, astronomer **Arthur Eddington** took on this challenge and traveled to the Isle of Principle off the coast of Africa to observe this phenomenon during an eclipse. After studying the photographic plates, it became clear that the stars did shift their apparent position in the gravitational field of the sun. Gravitational forces did bend the rays of the starlight by about 1.6 arc seconds, proving that light was a particle that was made up of mass and that Einstein's theories were holding true (Isaacson, pg. 257).

The figure below shows how the suns gravitational field pushes the apparent position of a star outward from its true position. We also see a similar effect due to gravitational lensing which also changes the apparent position of stars and galaxies due to massive objects throughout the universe.

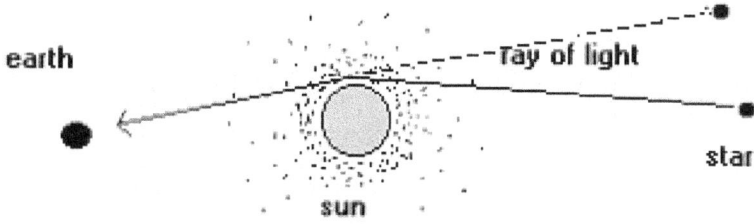

Apparent Position of a Star

Einstein developed his Principle of General Relativity, a unification of Newton's law of universal Gravity and his own Special Theory of Relativity. This theory would then go on to predict gravitational lensing, gravitational redshift, the precession of Mercury and many more scientific phenomenon discovered later in the 20th century. This principle was generalized in a mind experiment where passengers on a train moving along with uniform motion feel a sudden change in motion with the abrupt application of the brakes (Einstein, pg. 73). Einstein would go on to develop the **Equivalence Principle** from this showing that inertial mass that is produced by acceleration/ deceleration and gravitational mass in a gravitational field is essentially the same thing.

Gravitational Lensing

While writing for a science yearbook in 1907, Einstein noted a thought experiment of an observer in a closed windowless chamber that is accelerated upward. He compared the acceleration of this observer to another observer in a gravitational field showing that both felt pressed to the floor and that neither could distinguish the difference, thus formulating his "equivalence principle." This principle asserts that inertial mass and gravitational mass are equivalent, that both resistance to acceleration and gravitational effects are manifestations of inertio-gravitational fields.

In this same sort of scenario, an observer is in a chamber accelerating upward, again utilizing the "equivalence principle." A pinhole in one wall allows light from outside the chamber to shine on the opposite wall of the chamber. By the time light hits the opposite wall the chamber has moved causing the light to hit the wall a little closer to the floor, in effect **bending the light**. Since inertial effects are equivalent to gravitational effects, gravity too should cause the light to bend (Isaacson, Walter pg. 190). This is strangely similar to James Bradley's Star Aberration reviewed earlier with the "Tall Box Analogy".

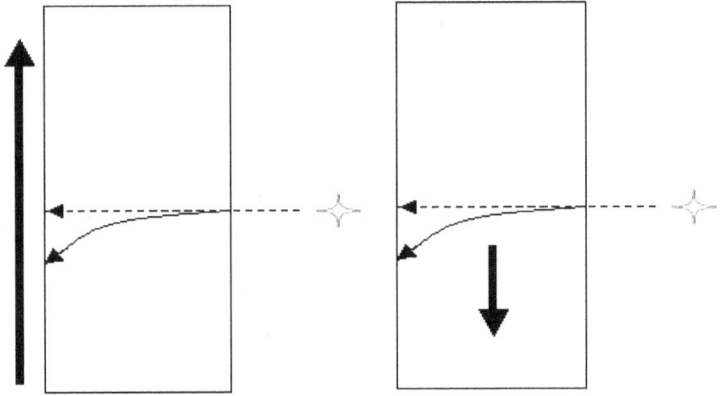

Equivalence Principle

Quasars, short for quasi-stellar radio source, were first discovered in the 1950's well after Einstein's theories were developed. Quasars are very powerful electromagnetic wave generators at the center of distant galaxies that are much brighter than normal galaxies. Quasars are the strongest sources of electromagnetic radiation in the universe, possibly generating energy from a super massive black hole at the center of its galaxy. Because of their great distances and brightness, quasars are ideal for studying the effects of gravitational fields on light. Einstein theorized that light would follow a curved path around massive objects due to the immense gravitational field. Today many stellar objects such as Einstein's Cross or Einstein's Rings occur due to the deformation of the light from a source such as a quasar through gravitational or cosmic lensing.

Gravitational lensing occurs when the light from a quasar or other very distant and bright source curves around an enormous object, such as a galaxy, that is between the quasar and the earth. In 1979 a "Double Quasar" named Q0957 was discovered with the first confirmed case of gravitational lensing. This double quasar was caused by the refraction like effects of intense gravitational fields bending the light of a single quasar into two observed quasars (Bord, pg. 498). More than 30 such gravitational lenses are known today including the **Einstein Cross** (G2237 + 0305 - on the cover of this book) where one quasar is refracted into four images of itself through gravitational lensing from a huge galaxy (Dekel, pg. 383).

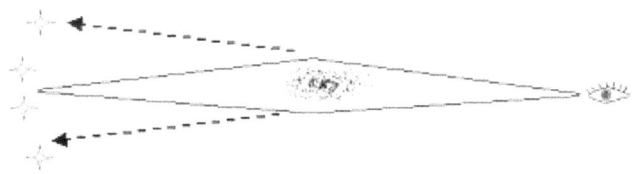

Gravitational Lensing

Gravitational Red Shift

The movement of the earth relative to a star or galaxy or the reverse, galaxy or star towards the earth produces a strange phenomenon known as **red shift**. Armand Fizeau, (inventor of the spinning toothed wheel experiment) described this spectacle in 1848 after noticing a shift towards red in the spectral lines of the light being viewed from stars moving away from the earth. Willem de Sitter had also noted this phenomenon in 1913 with the twin star Doppler Effect covered earlier. **Red shift** can be defined as a **stretching** in the wavelength of electromagnetic energy detected by an observer or detector compared with the original wavelength emitted by the star **moving away** from the observer. This expanded wavelength is also directly proportional to the drop in the frequency of the radiated light. The opposite is also true with the **compression** in the wavelength of light coming from another object **moving towards** the observer and/ or detector, commonly called **blue shift**.

Edwin Hubble's (1898-1953) earth-shattering discovery in 1929 suggests that **galactic Red Shift** in light from stars and galaxies increase in proportion to their distance from earth, this is known as Hubble's Law (Ditchburn, pg. 340). Studies show that 38 of 40 galaxies observed are rapidly moving away from each other. This omni-directional galactic expansion shows that galaxies furthest away are moving the fastest away from earth– some at 57,000 kilometers per second or 13 million miles per hour – 2% the speed of light (Bord, pg. 220). This process implies that some sort of repulsive force exists between the galaxies, similar to an expanding gas or substance that is pushing galaxies outward and away from each other – somewhat like bread rising. Gravity in effect should be pulling galaxies together in what ultimately would be the "Big Crunch". This possibly is why galaxies further away from each other are moving faster away from each other – gravitational forces are much weaker. Further research also shows that this expansion is accelerating additionally implying that an expansion agent also known as **Dark Energy** and **Dark Matter** (two different substances) exists in the universe. This substance is expanding between the galaxies and stars, forcing them outward, away from each other against their own gravitational attractions (Bord, pg. 493). Empty space may not be as empty as it may seem.

Gravity was now known not only to be able to bend light but even slow it down. A similar phenomenon known as **gravitational Red Shift** is noted in light emitted from very large stars, which is not related to movement but due to the immense gravitational fields of the star itself. The gravitational force of the star pulls at the light emitted from the star, slowing it down as if it had to climb up steep hill. This slowing effect on the light coming from the star causes the wavelengths of that light to become stretched, shifting the light toward the red end of the spectrum, called gravitational red shift. This Red Shift is only possible if the gravitational pull on the light emitted from the star is greater than the gravitational pull of the earth where the observer and detector are as the earth gravitational force has the opposite effect, speeding up the light (Ditchburn, pg. 315).

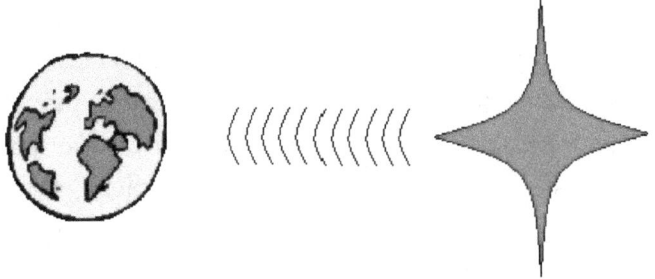

Gravitational Red Shift

Light reflected off the moon however, has the opposite effect, shifting towards the blue end, known as gravitational blue shift. This is only possible if the gravitational pull on the light reflected off the moon or any other object is less than the gravitational pull of the earth where the observer and detector are (Isaacson, pg. 148).

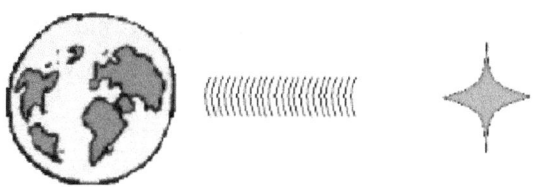

Gravitational Blue Shift

Highly accurate clocks onboard 24 **GPS satellites** circling the earth at 20,000 km (12,427 miles) altitude also are affected by Red Shift causing another phenomenon believed to be time dilation, a warping of space and time as predicted by special relativity (Bord, pg. 13). Gravitational Red Shift has a positive effect with GPS due to the weaker gravitational field of the satellite relative to the earth's surface. Time dilation however, has a negative effect due to the velocity of the orbiting satellite of 4 kilometer per second or 8,948 miles an hour relative to a stationary point on the earth, both tending to cancel each other out.

These GPS clocks must be synchronized to be within 50 billionths of a second in order to be capable of measuring positions on the ground within 4.5 meters (15 yards). These clocks must be corrected for "relativistic anomalies" due to gravitational red shift, otherwise they would run faster by more than 40,000 billionths of a second each day, changing the calculations of positions on the ground. Without these corrections, the clocks onboard the 24 satellites would be out to sync within a minute and a half – a matter of national security (Bartusiak, pg. 55). Einstein predicted that clocks would run slower in more intense gravitational fields (Isaacson, pg. 148). It is the earth's gravitational field that affects these clocks; therefore clocks at different altitudes should run at different rates having their frequencies shifted (gravitational redshift) due to gravitational differences. Two identical clocks at different altitudes will therefore run at different rates, with large enough difference to cause a variation in distance on the ground by as much as 13 km in a day.

Precession of Mercury

Another orbital phenomenon is the orbital precession of the planet Mercury. Mercury's orbit is an extremely elongated elliptical orbit approximately 36 million miles from the Sun that precesses slightly every orbit. Newtonian formulas explained this phenomenon by taking into effect the gravitational forces of all the other planets on Mercury including the effect from the Sun and its own rotation. Using these calculations, scientists calculated that Mercury's perihelion (the closet point of its orbit to the Sun) should shift or precess about 531 arc seconds per century. In reality, Mercury precesses about 574 arc seconds per century, 43 arc seconds more than the Newtonian predictions (Hobson, pg. 233).

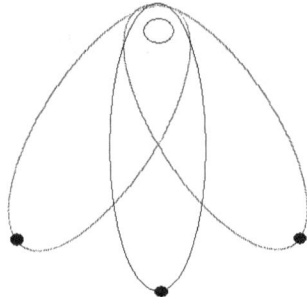

Mercury's Perihelion Shift

Newtonian formulas could not account for this change in motion. Something invisible was shifting the orbit ever so slightly, orbit by orbit. It has been suggested by some scientists that another planet, some call it **Vulcan**, may be in an orbit slightly closer to the Sun than Mercury, producing an additional gravitational force off center with the suns gravitational grip that could account for this discrepancy. Some other scientists have also theorized that there must be a certain amount of dust between the Sun and Mercury, which puts a drag on its motion and changes its orbit. No evidence however of dust or another planet has ever been found in this region. Einstein in 1915, calculated that the orbit of Mercury should precess by about 43 seconds of arc per century using the General Theory of Relativity, exactly what it does (Bartusiak, pg. 42).

Mercury's orbit ranges from 46 to 70 million km from the sun due to its elliptical orbit. Mercury is the fastest-moving planet in our solar system with a mean orbital velocity is 47.87 km per second with a minimum velocity of 38.86 km per second and a maximum velocity of 58.98 km per second, twice that of the earth. As Mercury changes velocity and enters into a stronger gravitational field during its Perihelion or closest encounter with the sun, its mass increases and its local time changes ever so slightly. These slight changes in mass and time due to the General Theory of Relativity help account for the missing 43 seconds of arc per century.

Tracking Venus and Mercury

If gravity of a large body in space such as a star could produce time dilation or a warping of space-time bending light rays, theoretically the sun could also produce this phenomenon. Working with astronomer Arthur Eddington in 1919 while viewing a solar eclipse, Einstein showed that the gravitational field of the sun actually did bend rays of light from a star. The position of the star was known but as its light rays passed close to its atmosphere, a form of Gravitational Redshift occurred, shifting its apparent position (Isaacson, pg. 257). The question arose that if the suns strong gravitational field affected light, could this same phenomenon be also evident with other electromagnetic energies?

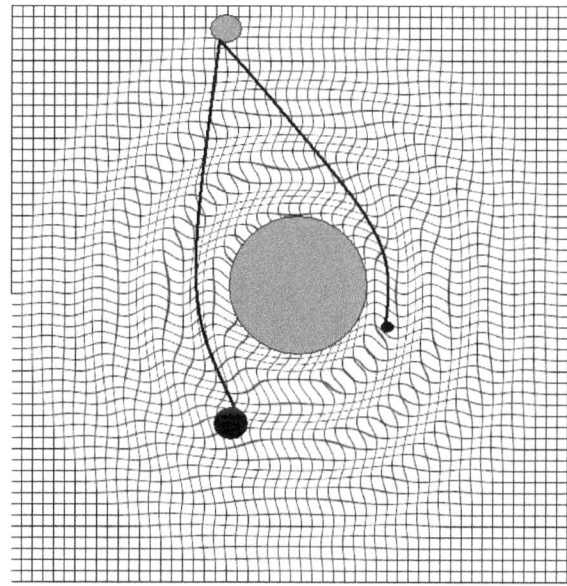

Radar Tracking of Mercury and Venus

In experiments carried out over several years beginning in 1964, Harvard Smithsonian astronomer **Irwin Shapiro** discovered a new variation of Gravitational Red Shift. Using **MIT**'s newly built Haystack Observatory located in Westford, MA, Shapiro started tracking and recording Mercury and Venus's positions when earth and each of these planets were on opposite sides of the sun. By bouncing high-frequency radar signals off the planets, Shapiro found that as the planets started to pass behind the sun, the return time of the radar signals increased by about one five-thousandth of a second. This lengthened the 23-minute round-trip time to Mercury and the 30 minute round trip time to Venus of these radar signals, simulating an increase in the distance of the planets orbits as if the radar beam had bent or "dipped" changing course, increasing the distance it traveled to and from the planets (Bartusiak, pg. 51).

Cosmic Background Radiation

In 1964 while working on ultra-sensitive cryogenic microwave receivers, Bell Laboratory researchers Arno Penzias and Robert Wilson came across an unexplained microwave radio noise (Bartusiak, pg. 63). This pesky radio noise was isotopic, meaning it came from all directions, far weaker than the radiation given off by the Milky Way. They first assumed that their instrument was picking up interference by local radio and TV station noise from New York City but noticed that it came from all directions averaging in the microwave range of approximately one millimeter. The next assumption was that a pigeon dropping in the radio antenna's large horn was causing the microwave radio noise. After cleaning up the pigeon mess, the noise still frustratingly remained (Bord, pg. 304). Further research showed that this radiation was from so far away that it must have been generated over 15 billion years ago, a half a million years after the "**Big Bang**", before the formation of the planets when the universe was filled with a red-hot mist of plasma.

Optical radiation began to be generated as the first atoms began to form; the "primordial fog" was then lifted and the universe became relatively transparent, beaming with light (Bartusiak, pg. 226). Prior to the formation of atoms, the universe was probably amassed with only the building blocks of atoms: protons, electrons and neutrinos. Research on this newly found phenomenon shows that galaxies are all rushing away from each other at very high velocities and seems to show that the universe is constantly expanding (Ditchburn, pg. 340). It is believed that as the universe expanded, the hot plasma evenly distributed through the universe cooled. The light that was being generated at that time, mainly in the optical and infrared spectrum, became stretched, increasing its wavelength and lowering its frequency down from the optical range, down into the microwave range. While viewing the universe today with a traditional optical telescope, the background of space is now totally black. With a radio telescope however, there still remains a weak glow in all directions in the microwave radio frequency range of 160.2 GHz; this microwave radiation now being detected is scientifically known as **Cosmic Microwave Background Radiation or CMBR.**

Could this energy at the edge of our known universe be the remnants of the plasma that was generated during the Big Bang? Could this background noise actually be the energy of the material that all particles condensed from? If so, this relic that we see is the way it was 15 billion ago, which has by now has condensed into stars and galaxies, a universe we can never really see.

Dark Energy and Dark Matter

As previously mentioned, **Dark Energy**, the hypothetical expansion agent, possibly causes the rapid dispersion of our universe. As particles of matter gather great distance between themselves, gravitational attraction between them begins to wane, allowing this expansion agent to more rapidly separate particles from each other, accelerating them exponentially (Hawking, 1988, pg. 48). The universe is believed to be composed of 73% of this mysterious **Dark Energy**, 23% of an unknown type of **Dark Matter**, and 4% of ordinary matter. Through understanding "Hubble's Constant" and calculating the known visible mass of galaxies, the universe should be expanding at a specific rate, but it is not. Through measurements of redshift in the light emitted from these galaxies, the universe is expanding much faster than first believed. The only way to explain this is to calculate in the missing mass (Glendenning, pg. 23).

The measurement from the Wilkinson Microwave Anisotropy Probe (WMAP), which was launched in 2001, has played a primary in developing today's Standard Model of Cosmology. WMAP, through the study of Cosmic Background Radiation has also determined that the cosmos is relatively flat and 13.7 billion years old (Glendenning, pg. 64).

This Background Radiation was determined to be homogenous or proportionally the same throughout the universe as a "smooth sea" of energy even though the stars and galaxies are clumped together in large masses with "nothing" in between them (Bartusiak, pg. 63). This energy seems to act as a hot gas, expanding outward, pushing everything around it away. As larger distances separate galaxies, the gravitational pull that attracts them diminishes, allowing this expanding "gas" more leverage, accelerating these masses faster and further apart. This was Einstein's **Cosmological Constant** which represented a repulsive force in the universe, counteracting gravitational forces that attract bodies throughout the universe (Isaacson, pg. 353). This constant was developed in order to maintain a balance in the universe, preventing it from collapse – a static view. After Hubble's discovery that indicated an expanding universe, this equation was modified, allowing for Hubble's Law (Ditchburn, pg. 340).

In the 1930's, Cal Tech professor <u>**Fritz Zwicky**</u> was first to conceive of **Dark Matter** in an effort to account for the high velocity of stars in the outer fringes of our own galaxy. Theoretically, these stars would spin off into space due to centrifugal force, but an unseen mass needed to be added, as gravity due to the visible mass of luminous stars was not enough. Dark Matter can also be used to account for gravitational lensing that curves light rays passing through the immense gravitational fields of galaxies. There has to be much more mass exerting gravitational forces than was seen visibly. The majority of Dark Matter is believed to be non-baryon, meaning that it does not have the atomic structure of visible mass such as the hydrogen atom and may not be able to interact electromagnetically with discernible mass other than gravitational forces (Glendenning, pg. 49). Dark Matter would evidently not be associated with large **black holes** in much of the universe, as gravitational lensing would indicate their presence, however, the existence of miniature black holes permeating the galaxies proportionately dispersed, is still a possibility.

The Muon Life Cycle

Muons are highly unstable subatomic particles with a life span of 2.2 micro seconds (.0000022 of a second) and are only produced under intense energies. Muons like electrons are classified as leptons and have a mass which is about 200 times that of the electron (Bord, pg. 451). Outside of nuclear events, muons are only naturally created by cosmic radiation composed of protons traveling at near light speed from deep space impacting air molecules at the edge of space. Given their short life span, muons should never reach the ground, decaying into an electron, an electron-antineutrino, and a muon-neutrino high up in the atmosphere (Bartusiak, pg. 34). Mathematical calculations predict that less than one muon out of a million should ever hit the earth's surface. Experiments however show that about 50,000 out of a million muons created in this exchange hit the ground and that 10,000 muons hit every square meter of the earth every minute of every day – how could this be?

Only a relativistic resolution can be found. It is believed that muons traveling at .98 times the speed of light undergo time dilation where local time on the muon is much slower or length contraction occurs which allows for them to exist long enough, 2.3 times its life span, to hit the ground (Bord, pg. 452).

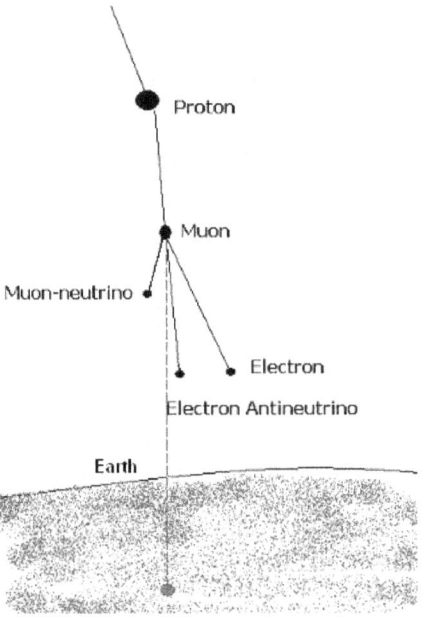

Muon Life Cycle

Review of History

Mankind has always wondered about the universe. Where did we come from and what are we made of? The questions about light, its composition and how it works has always intrigued us, even back to the 300 BCE's, in the days of Plato, Aristotle and Socrates. In those days, light was considered the fifth element; the "Luminiferous Aether", the stuff stars were made of. It wasn't until Galileo in the early 1600's that the question of light's speed was tested with the use of covered lanterns. All that he could determine at that time was that the speed of light was at least 10 times faster than the speed of sound. Not too much later in the late 1600's however, Ole Rømer determined through the study of Jupiter's moons that the speed of light was 299,792 km/sec or 186,290 miles/sec, becoming the cornerstone of quantum physics, the study of atomic matter.

Various theories of what light consisted of, slowly developed, including René Descartes undulation theory, otherwise known as wave theory in the early 1600's and Sir Isaac Newton's corpuscular theory or particle theory in the early 1700's. Newton's particle theory however did not stand very long as Thomas Young in the late 1700's used the "Double Slit" experiment where light created interference patterns, proving that light was composed of waves. Studies showed that light was made of both longitudinal waves and transverse waves complicating the aether theory of the day. This suggested that the aether, the medium that light traveled through, must be created of a solid substance, which would resist the motions of the planets and the stars and did not seem to make sense.

In the 1700's, James Bradley became the first to prove the Copernican heliocentric view of the universe. It became evident through his studies that the earth moved very rapidly though space in its orbit around the sun. These studies of star aberration, or the apparent shift in a stars position, showed different displacements of the stars over time in the northern region around the North Star and south in the equatorial region. This lead to his calculation of the speed of light which was very close to Ole Rømer's calculations 100 years earlier. This knowledge was reinforced by Fizeau's and Foucault's experiments showing lights speed within a couple thousand miles per second.

Later on in the late 1700's François Arago tried to show that light traveled at different speeds due to gravitational forces of stars and the movement of the stars. His experiment, using a prism and a telescope was set up to detect these various speeds using Newton's refraction theory. The test failed; at least in showing that there was no variation in the refraction of light. This experiment added confusion to the mix by indicating that the speed of light from large stars and small stars, and that of moving stars in relation to the movement of the earth was the all same.

The aether or medium, through which light moved, at this time, was thought to be dragged along with the earth through space. Star aberration however disproved this, showing that the aberration was due to lights displacement by the earth's movement and was not dragged along by the atmosphere. French physicist Augustin Fresnel would then develop the partial aether drag theory, suggesting that light traveling through air would not be dragged along, but that as the refraction index of a substance increased, so also would the drag effect increase. Light would therefore travel unimpeded through the air and be uninfluenced by the movement of the earth. Once light enters into glass or some other medium of higher refractive indexes it would therefore be constrained or locked into the new medium and travel at one velocity with that medium.

George Biddell Airy later in the 1800's tried to detect the aether using a kind of water filled telescope. The theory was that light traveled slower through water and that while observing star aberration due to earth's movement through space, the aberration angle would change. This also failed; indicating that the aether or medium which light traveled did not exist, adding perplexity to this study. In the late 1800's Albert Michelson and Edward Morley went on to disprove the existence of the aether with their interferometer experiment.

Willem de Sitter, early in the 1900's discovered that light traveled the same speed in his studies of twin stars; much like Arago's experiments in the 1700's. His discoveries showed the Doppler Effect evident in the shifting of the light spectrum towards the red or blue due to the movement of these stars. Although the speed of light remained the same to the observer, the frequency of the light viewed changed proportional to the speed and direction the star was moving in reference to the observer.

These experiments confirmed to scientists of that day that there was no aether. Not only did these tests show that there was no aether, they had also discovered what is now called "Law of the Constancy of the Speed of Light." This measurement of the speed of light is the same no matter what the observer's velocity is what velocity the emitter is, either moving towards or away from each other expressed as "c" and is measured at 299,853 km/sec. or 186,328 miles/sec.

Hedrick Lorentz would then try to rectify this strange problem by developing the "Lorentz Contraction" theory. This theory showed that the movement of an object causes length contraction of that object in the direction of its movement, therefore causing the speed of light to remain constant no matter what the movement of the observer or object being observed. This is one explanation in how the Muon can reach the earth at high speeds even though its life cycle supposedly would prevent it. The "Lorentz Contraction" due to its velocity at near light speed, changes the space-time environment of the Muon, allowing it to reach the ground.

Albert Einstein would then build on this to develop his "Theory of Relativity". Einstein theorized that movement changed time and space and that the faster an object moves the more time and space around it would change. Gravity also would have a dramatic effect on space and time causing light waves to bend around stars and for time to vary in different gravitational fields. This explains the constancy of the speed of light and the shift in the light spectrum due to gravitational fields and light from objects at different velocities.

Modern day studies seem to verify Einstein's theories beginning with studies on the precession of Mercury. Studies of starlight being bent by the suns gravity, gravitational lensing due to the intense gravitational fields of galaxies also validates this theory. Galactic redshift and gravitational redshift is seen in the GPS transmission or radio waves and even the tracking of Venus and Mercury with radar. It seems that no matter how you look at it, everything seems to indicate that Einstein was correct.

Edwin Hubble's discovery in 1929 showed that while studying galactic Red Shift that redshift in light from stars and galaxies increase in proportion to their distance from earth. This developed into what is now known as Hubble's Law, expressed by the equation $v = h_0 d$ where v is the velocity of a galaxy. The further we look out into the universe, the faster galaxies and stars accelerate away from each other, indicating that these galaxies and stars were moving out from a single point in space over a long period of time.

This discovery lead to the "Big Bang Theory" as scientist began to work back in time to a point where this rapidly expanding universe theoretically began, to a one point in space and time called the "singularity" over 13 billion years ago. In 1964, Arno Penzias and Robert Wilson would discover Cosmic Microwave Background Radiation (CMBR), evidently a relic of the "Big Bang" dispersed evenly across the sky.

Studies from observing these stars and galaxies generated new ideas, one being Dark Energy. What was the force that acted as a dispersing agent, pushing the galaxies and stars away from each other in deep space; an energy that was repelling against the forces of gravity, forces that should pull everything together in what eventually would be the "Big Crunch"? Another recent observation also suggests the existence of Dark Matter, an invisible mass that would generate enough gravity to keep the rapidly spinning galaxies from flying apart. Existing "known" mass in galaxies such as our own Milky Way, does not have the amount of mass needed to keep stars on its outer edge, moving at great velocities, from flying off into space.

CHAPTER III

RELATIVITY

The "Lorentz Transformation" Hypothesis

The failure of Michelson and Morley to detect the aether was puzzling. Soon after the Interferometer experiment, Irishman George F. Fitzgerald (1851-1901) professor of experimental philosophy suggested that the negative results of the experiment might have been due to the contraction of the interferometer in the direction of movement. Hendrik Lorentz built on this idea developing the contraction hypothesis based on his Electron Theory stating that any material body is contracted in the direction of its motion. This contraction would then change by a factor relative to the speed of light (Einstein, pg. 10).

Lorentz's contraction theory would soon serve as the foundation for the mathematics of Einstein's theory of relativity. Einstein developed on this hypothesis making some substantial changes. Einstein modified the Contraction Theory showing that these changes due to motion do not actually warp the body in motion, as Fitzgerald and Lorentz had originally suggested, but instead warped or made changes in space and time around that moving body (Ditchburn, pg. 318).

Lorentz's transformations, which he introduced in 1904, formed the basis of Einstein's Special Theory of Relativity. They describe the increase of mass, the shortening of length and the time dilation of a body moving at speeds close to the velocity of light. This may all seem out of the norm as no one or nothing man has yet made even travels close to these speeds. However, this does affect our Earth and us on the Earth. The Earth moving in its orbit around the Sun moves at about 30,000 meters/sec (30 km/sec or 18.5 mile/sec). The contraction of space caused by the high velocity of Earth through space would contract the diameter of Earth by about 6 cm (2.5 inches). This small change may account for Michelson and Morley's negative result by making the source of light and the mirror draw closer together when the system is moving lengthwise nullifying any change due to the earth's rotation and velocity through space.

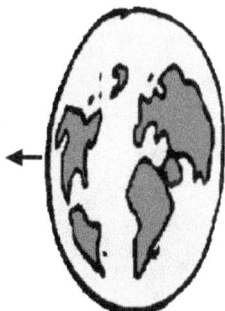

Lorentz Transformation of Time and Space

Lorentz Transformation of Time and Space

In three-dimensional space, the Moving System moves with velocity v in the x direction with respect to the Fixed System reference; the moving X' is contracted.

"z" represents height, "y" represents width and "x" represents length of an object. Here, measurements are compared to a fixed system relative to a moving system. Note – Length "x" is not =.

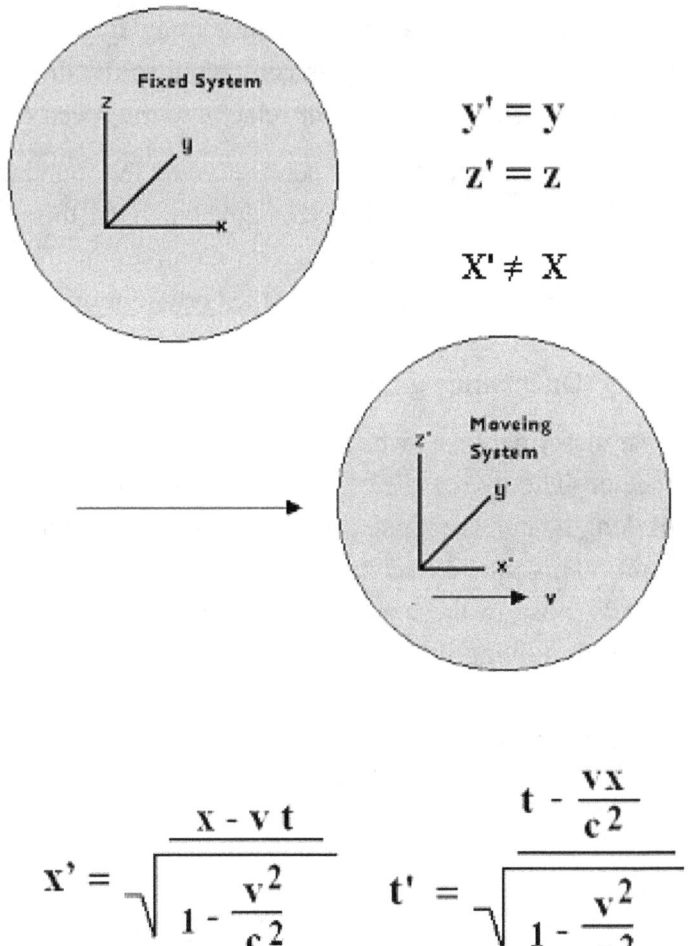

$$y' = y$$
$$z' = z$$
$$X' \neq X$$

$$x' = \frac{X - vt}{\sqrt{1 - \frac{v^2}{c^2}}} \qquad t' = \frac{t - \frac{vx}{c^2}}{\sqrt{1 - \frac{v^2}{c^2}}}$$

Lorentz Transformations

Lorentz used an arcsine or inverse sine formula to express these transformations.

$$= \neg \sqrt{1 - a^2/b^2}$$

This formula divides the square of two values (such as two velocities) in order to show the ratio between these two values in a range from 0 to 1 (A). This ratio is then subtracted from one to return an inverse nonlinear value that cannot be greater than 1 (B). This value is then factored with other values that determine the change in length, time or mass variable that changes according to velocity (C). The number one or maximum values is then substituted with the value of the speed of light, 186,000 miles per second or 300,000 kilometers per second; the percentage can also be used such as .8 to express 80% light speed.

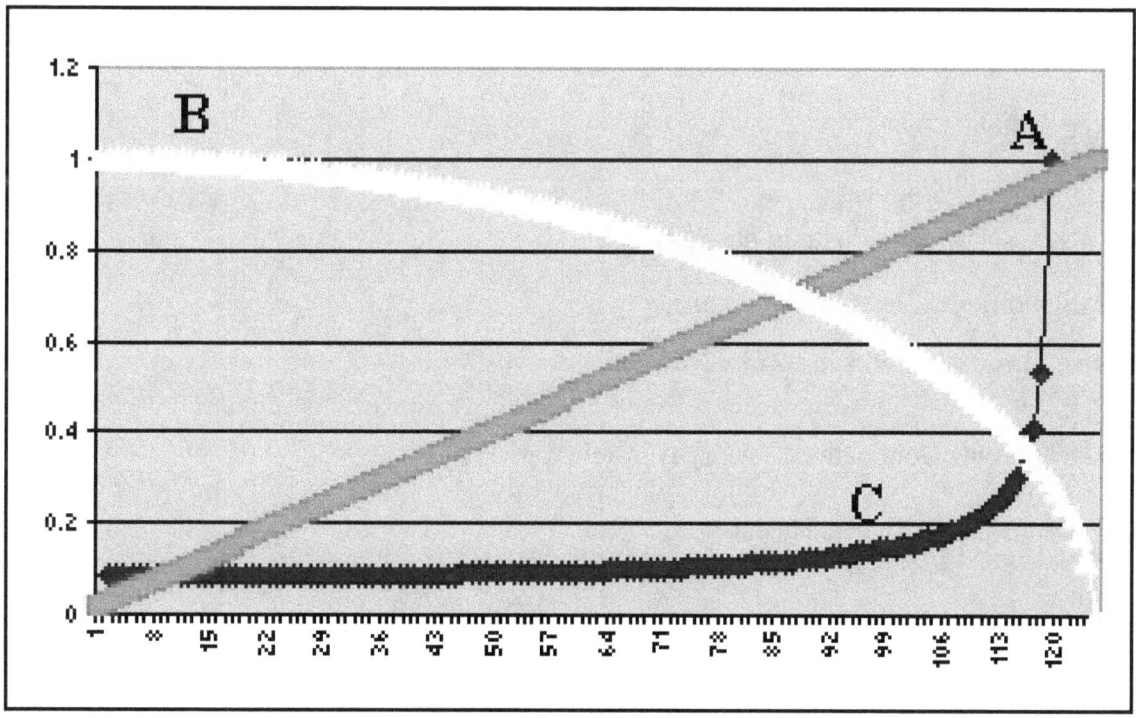

Formula Variations

Example:

Fixed System

v = velocity = 0

l = length = 100 ft.

t = time = 10 seconds

c = speed of light

m = mass = 10 Newton's

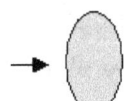

If velocity v = .5 c then

Length of contraction = .866 x 100 = 86 ft.

Time dilation = 1.15 x t = 11.5 seconds

Relative Mass = 1.15

m = 11.5 Newton's

Moving System

If velocity v = .999 c then

Length of contraction = .044 x 1 00= 4.5ft

Time dilation = 22.366 x t = 3.7 minutes

Relative Mass = 22.366 x m =223.7 Newton's

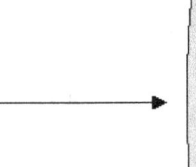

The "Lorentz Contraction" Theory (Einstein, pg. 53)

The Lorentz Time Dilation Theory

$$l' = l \sqrt{1 - \frac{v^2}{c^2}}$$

Contraction of Space

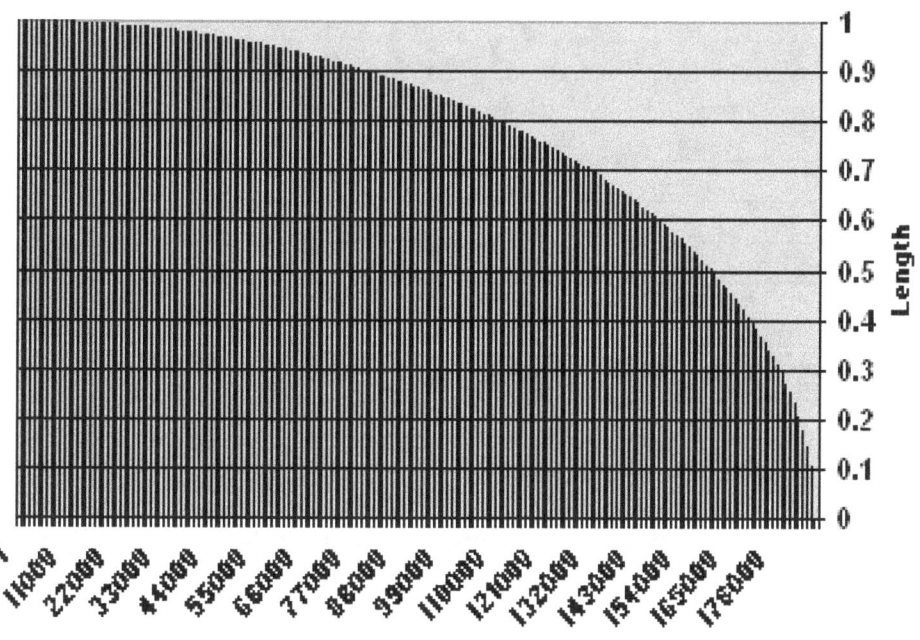

Length Contraction

The Lorentz Time Dilation Theory

$$t' = \frac{t}{\sqrt{1 - \frac{v^2}{c^2}}}$$

Dilation of Time

Time Compared with the Velocity of an Object

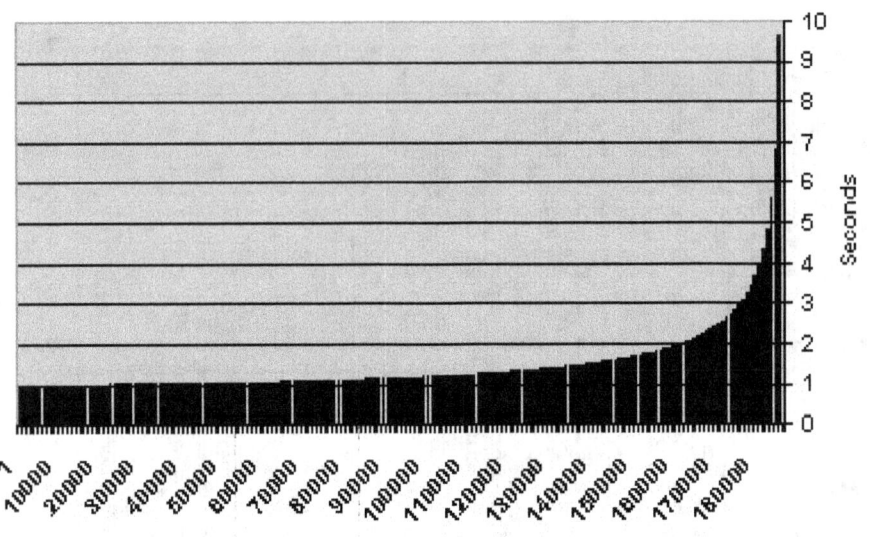

Velocity

Time Dilation

Relativistic Mass Formula

$$m = \frac{m_0}{\sqrt{1 - \frac{v^2}{c^2}}}$$

Mass Change in Velocity

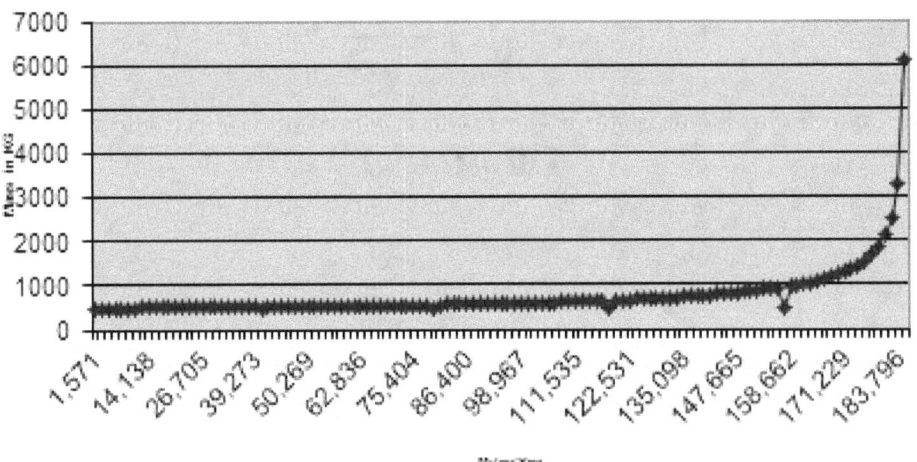

Relativistic Mass

Einstein's Remedy

As we saw earlier, Einstein's remedy for this dilemma in the early twentieth century was to alter time and space, building on Hedrick Lorentz's Contraction Theory. As discussed previously, if a baseball is flying at 100 mph and is measured by a radar gun that is moving in a vehicle towards it at 50 mph, the reading on the radar gun would read 150 mph. The only way to keep measuring the velocity of the baseball at 100 mph instead of 150 mph is to alter "time and space." We would therefore either have to change the length of the measuring rod, as Einstein would say it, or change the rate of the clock or both. In this case we would have to change the way the radar gun calculates the rate of change in the baseballs position.

This of course does not affect our baseball game; baseball speeds just aren't fast enough. This concept is however very important to understand as we venture further into the future and develop new technologies. If we ever do develop these hyper speeds, what would happen to us at these velocities? According to Einstein for example, if the measuring rod on the moving object was a yard, it would then have to be perceived to be shorter by a stationary observer (Contraction of Space) and one second to the stationary observer would be less than one second on a clock on the moving object (Dilation of Time). Of course inside that moving system the yard would still be perceived as one yard and one second perceived as one second; no change at all would be perceived in space and time by an observer moving with that system (local space and time). In that way, to a moving observer, no matter what their velocity towards or away from the light source, their readings while detecting the speed of light will always read 300,000 km/sec or 186,000 miles/sec.

In theory, velocity changes space and time. The faster the velocity of an observer, the slower he moves forward into time compared to a stationary observer. Below is an illustration of Einstein's Co-ordinate Systems A and B. System "A" represents a stationary system and System "B" represents the moving system. The embankment - (System A) in Einstein's Train analogy symbolizes the stationary system while the train represents the moving system (System B) (Isaacson, pg. 123). The height or vertical plane of three-dimensional space in both systems is represented by y where $y' = y$. The horizontal aspect of width in both systems is represented by z where $z' = z$. The horizontal plane that corresponds to the fore and aft movement in three-dimensional space in both systems is represented by x where $x' = x -$ velocity multiplied by time or $x - vt$. The length of an object is shorter or compressed in the direction of movement by a magnitude the faster an object moves as seem in length x' relative to x.

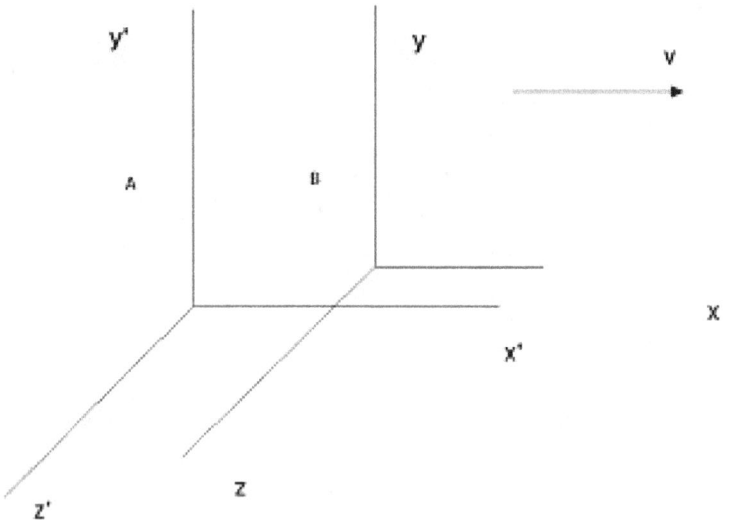

Co-ordinate Systems – Contraction of Space

Illustrated below is the variation in time (Local Time) between three moving systems traveling at three different velocities. System "A" is traveling at 1000 mps (miles per second relative to a stationary observer) where 1 second is equal to 1.000014 seconds. System "B" is traveling at 170,000 mps where 1 second is equal to 2.464495 seconds. System "C" is traveling at 180,000 mps where 1 second is equal to 3. 969143 seconds. Time moves slower in a moving system than a stationary system, therefore theoretically, the faster an object moves the slower a clock ticks. At a very high velocity, say on a spaceship, this time dilation would make it possible for passengers to travel further into the future while aging very slowly compared to people on a stationary object (Born, pg. 317-320). In other words, everyone on earth has aged much more than those in the spaceship

Seconds	Velocity
1.000014	1,000 mps
2.464495	170,000 mps
3.969143	180,000 mps

System Velocities

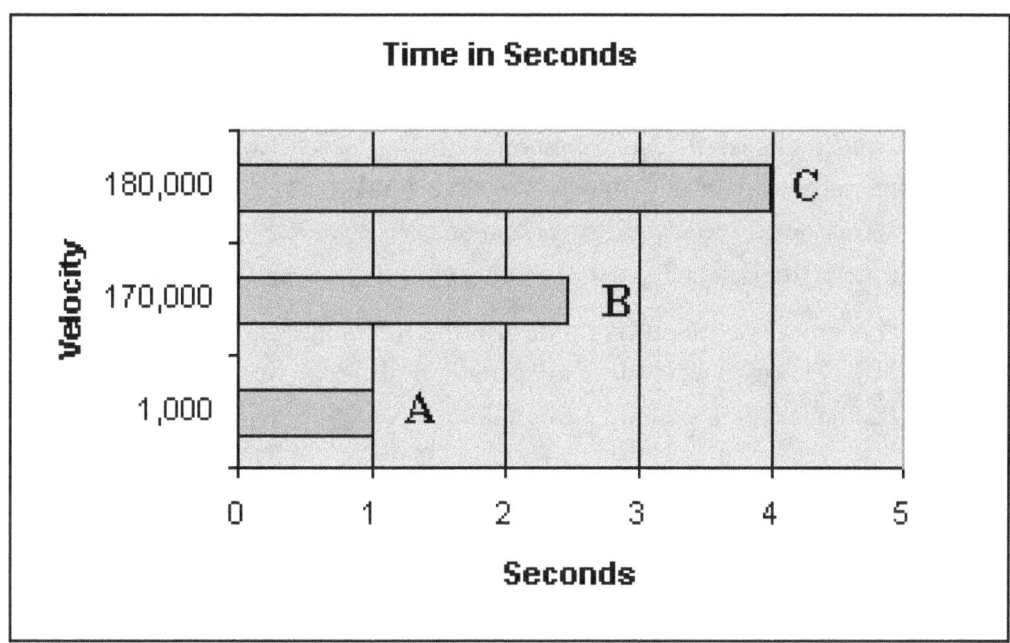

Co-ordinate Systems – Dilation of Time

Blast Off

To test this theory, let us perform what Einstein called, a thought experiment. When? It is the year 2050 and technology has brought us to a turning point. Using a now safe Cold Fusion nuclear power plant and the latest version of the Ionic Thrust Engine (ITE), we are now ready for a venture never before possible. Assembled in orbit around the earth, a five-stage space vehicle is readied for launch with five crewmembers on board. Its fuel is water, accelerated in vapor form to approximately the speed of light by ionizing droplets of the water and accelerating them though a magnetic field. Each stage will "burn" for approximately twenty-four days straight, maintaining an acceleration of 98 ft. /sec^2 and a constant force on the crew of 3g's (three times Earth's gravitational force) for the entire burn. If you would experience this in a car, this would be an acceleration of approximately 0-66 miles per/hour in one second maintained for twenty-four days straight, a little rough on the human body.

The crewmembers are strapped in and hooked up to devices that provide nutrients and also supply and remove fluids and toxins from their bodies over the next twenty-four days. After blastoff, the crew monitors the spacecraft for a few hours, and then is put into a deep sleep. After these twenty-four days are up, the crew are gently awakened and allowed to adjust to the environment. Then on the twenty-fifth day, engines are shut down and the acceleration stops – the crew are weightless. Crewmembers unbuckle themselves and enjoy this release of pressures on their bodies and are given a day to recuperate.

At the end of their one-day break, one crewmember enters the observation cabin on the first stage booster. The booster is then separated from the space ship; the crew is readied, and then blasts off again leaving Stage One and one observer behind. Like the first stage, this ship also accelerates to 3 g's for twenty-four days straight and the crewmembers again sleep away the time until awakened for the next brief rest. Like previously done, the second stage and one observer are left behind and Stage Three blasts off, accelerating on at 3 g's for twenty-four more days off into space.

This process goes on five times till the fifth stage and its lone observer shut down the engines and cruise weightless in space in a different world. Where are they and how far and fast have they traveled? Let's find out.

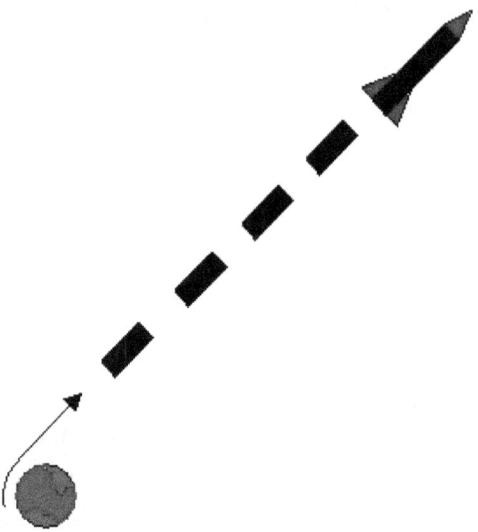

Five Stage Rocket to Light Speed

First burn - How far did the crew travel and how fast are they traveling having maintained 96 ft./sec/sec or 3 g's for twenty-four days straight? Remember, this was 0-66 miles per hour in one second. After one minute or 60 second later, the crew would be accelerated to 3,927 miles per hour and after one hour, 235,636 miles per hour. When twenty-four hours or one day has elapsed, the space ship will be traveling at 5,655,272 miles per hour or 1,570 miles per second. When the first twenty-four days have passed and the crew is on their first well needed break, they will be traveling at 37,701 miles per second, one fifth the speed of light.

Second Burn - This next stage blasts off and accelerates for twenty-four more days up to 75,403 miles per second. Third through Fifth Burn - this process continues until the fifth stage completes its burn traveling at a calculated velocity of 188,509 miles per second, faster than the speed of light in reference to the Earth. A total of 121 days have now passed and the spacecraft has traveled more than 60 billion miles.

Now that we have gone this far, it's time to return back to earth. Each stage of this ship adrift in space retained half of its fuel (water) for the return home. The fifth stage and its resident take one day to recuperate and then turn the ship around, firing up the engines for the long return trip back home. Blasting off back towards earth, the fifth stage returns back toward the direction it came from, with a rendezvous of course in store with Stage Four floating aimlessly in the deepest parts of the universe.

Like all of the other separation launches of each stage, Stage Five also accelerates up to and maintains an acceleration of 98 ft. /sec^2 or 3 g's. This acceleration however is back in the direction of which Stage Five came from and is in fact a deceleration, facing in the other direction. The effect on the ship and its passenger is however exactly the same, being pinned up against the seat for 24 days straight. What we are doing is actually slowing down Stage Five to Stage Fours velocity and catching back up to it to rejoin the capsule and crew. This will go on for four more link ups as the space ship returns back to earth, taking another 121 days to complete. Each launch decelerates the ship down to the velocity of the prior stage until the ship returns back to earth's orbit whole again.

If this is hard to understand, let's put it this way. When Stage Five blasts off and leaves Stage Four behind, it will be traveling after 24 days at 37,701 miles per second away from Stage Four. In space however, this is all relative. Who is to say that Stage Five is moving away from Stage Four? Stage Four could also be viewed as moving away from Stage Five at the same velocity with Stage Five being stationary in space (after it stops accelerating of course). If this were the case, Stage Five would have to use its rocket engines to catch up the Stage Four. This is in fact what we are doing on the return back to earth. With no real reference of what is stationary space, anything not accelerating or decelerating can be considered as stationary. Each stage viewed from another will be traveling at a different velocity – see Relative Velocity chart.

Now, according to calculations, the observer on Stage Five before his return on day 121 was theoretically traveling at 188,509 miles per second relative to the earth. Stage Five would be therefore traveling 2,218 miles per second or almost 8 million miles per hour faster than the speed of light. Can this be?

58

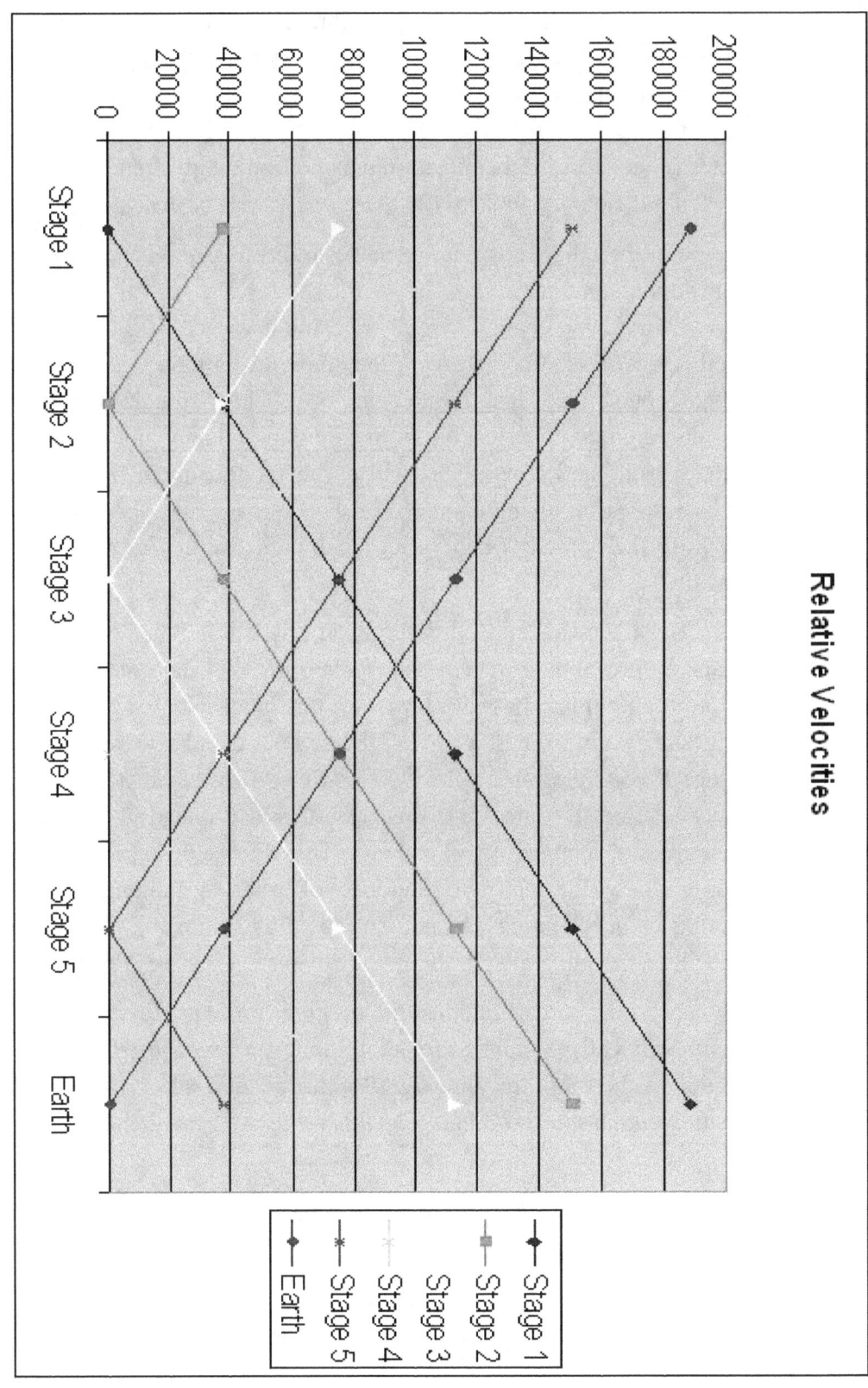

Relative Velocities

59

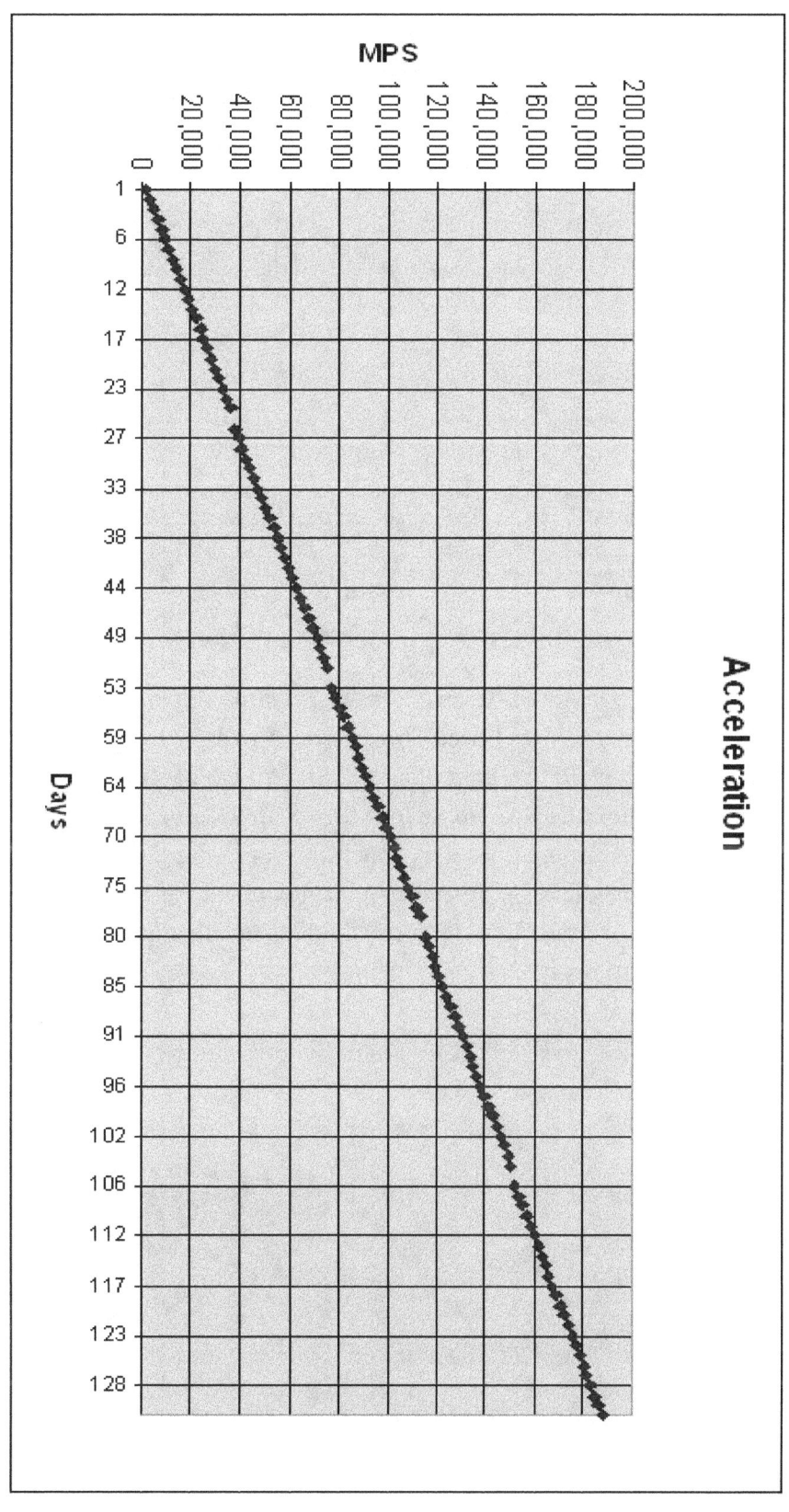

Theoretically nothing can go faster than light. However, if the first stage of a two stage rocket reached velocities close to the speed of light say .95 times the speed of light, and launched the second stage up to the same velocity away from Stage One, you would have this scenario.

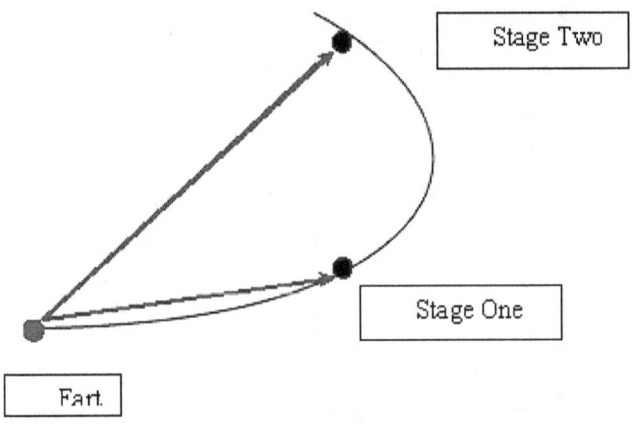

Warping of Space

Einstein's Theory of Relativity would "warp" or bend space in order to prevent Stage Two from going "Warp Speed" or the speed of light. From the Earth's perspective, if Stage One was at .95 the light of speed, Stage Two could not go very much faster, say .98 the speed of light and would have to travel in a curve. The Lorentz Contraction Theory introduced by Einstein does very much the same thing by reducing the rate of change in distance exponentially. This rate of change occurs the faster an object moves away from or towards an observer until the velocity does not change at all, preventing an object from reaching "Warp Speed."

61

Length Compaired with the Velocity of an Object

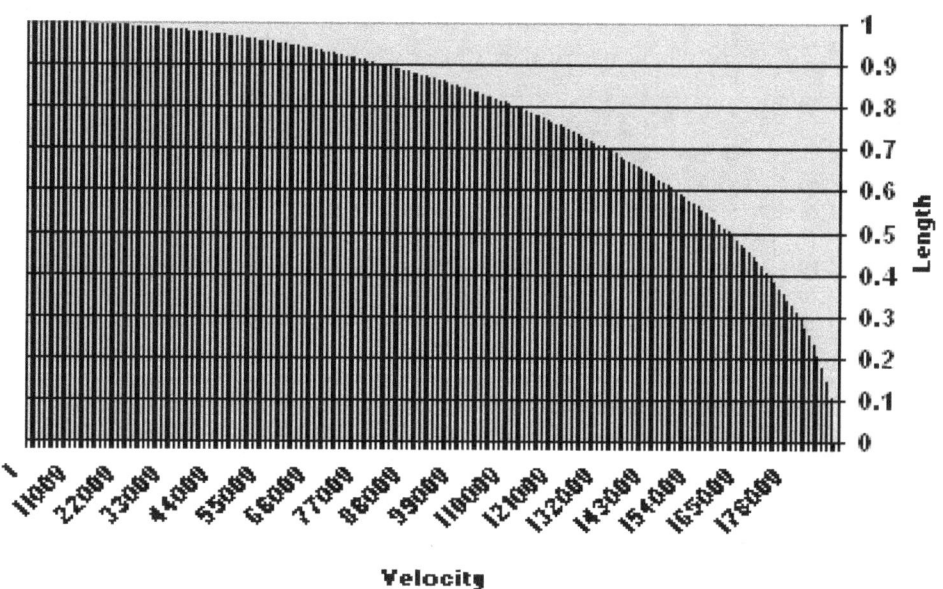

Contraction of Space

From the earth's perspective (see chart on following page), with each stage floating in space moving at different velocities relative to the earth, lengths will contract more and more. For example, with a ten-stage spaceship, stage one will be viewed as having very little change. Stage five however will have a contraction of approximately 80% earth's measurements and stage 10 with length contraction around 1% earths measurements. From stage 5's perspective, both the earths and stage 10's lengths will contract to about 80%. Stage 10 will view length measurements of other stages opposite that of earth with earths lengths at 1% and stage 5 at 80% that of stage 10's measurements.

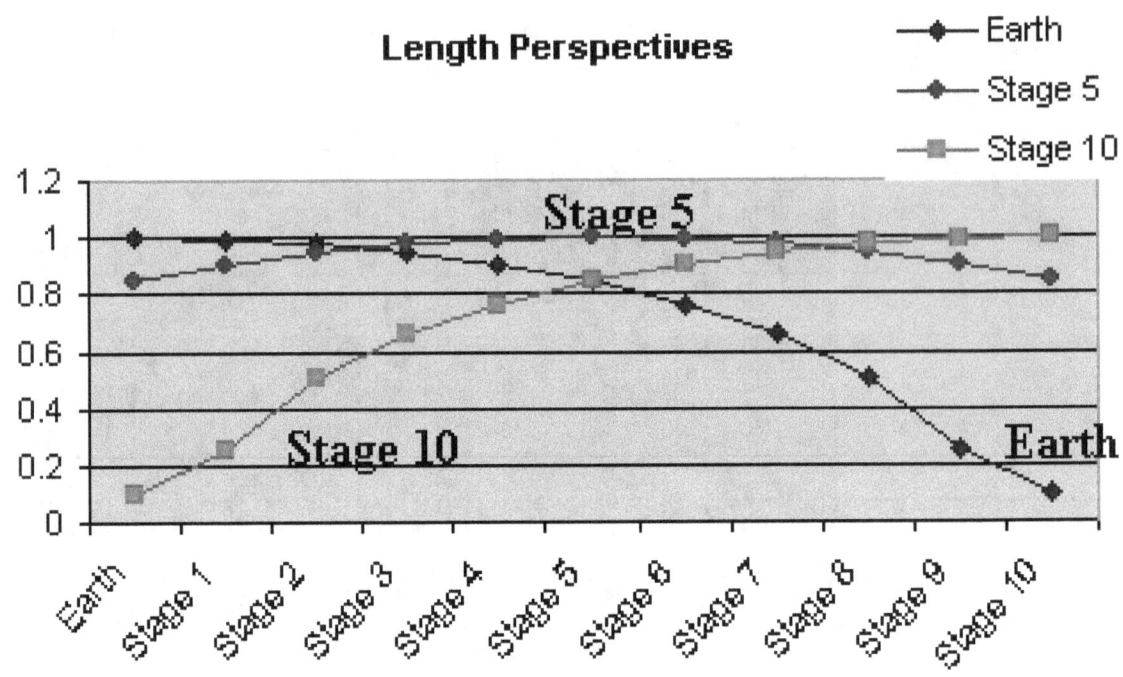

Length Perspectives

In another perspective, space would be compressed in front of a high-speed object similar to a bow wave in front of a boat and be stretched or dilated behind the object. Einstein had suggested that it was space – time that warped, not the moving objects.

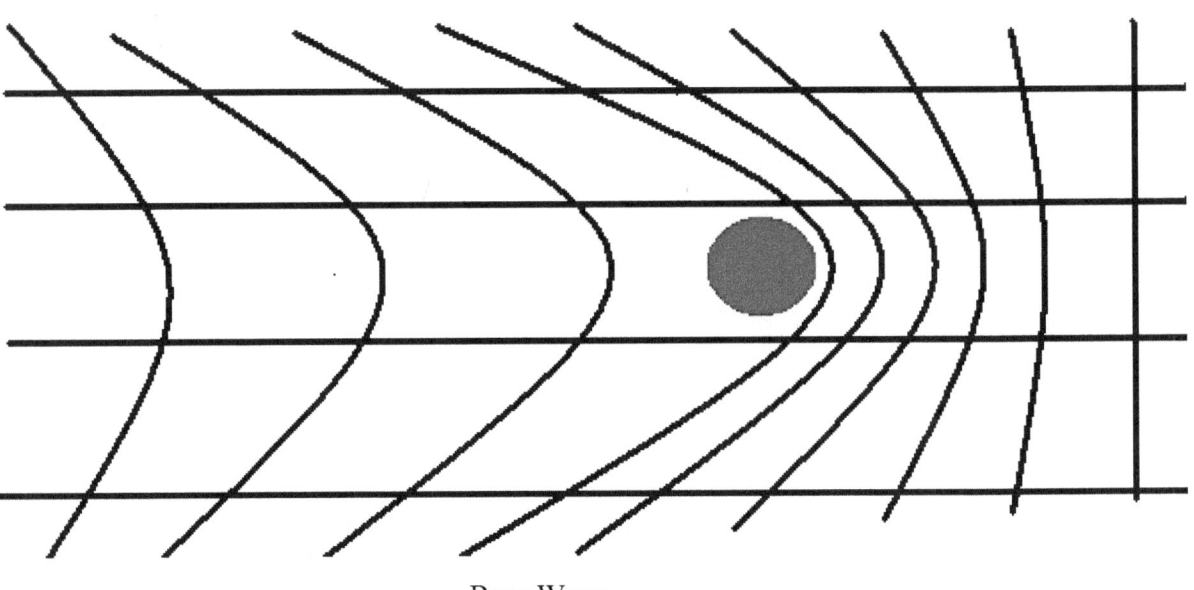

Bow Wave

If time is changed, the distance traveled must also change. Therefore, another perspective is to "warp" time in order to prevent Stage Two from going "Warp Speed." According to Einstein, the Dilation of Time would exponentially effect time the faster an object moves away or towards an observer.

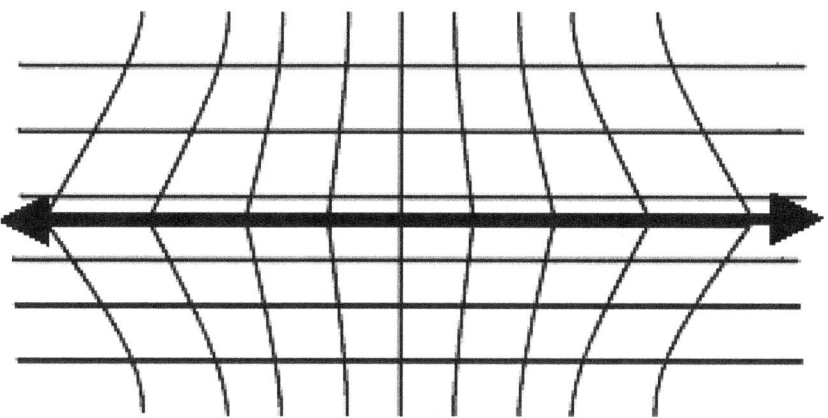

Dilation of Time

Going back to our rocket mind experiment, while viewing this rocketry from earth, Stage 5 hits terminal velocity at day 120, reaching a calculated velocity of 186,291 miles per second relative to the earth. At day 121, if it can continue to accelerate above light speed, it will reach a velocity of 188,509 miles per second, 8 million miles per hour faster than the speed of light. However, while viewing Stage 5 from Stage 1's observer's perspective, Stage 5 can accelerate on to day 146 where relative to Stage 1 it will hit terminal velocity at the speed of light. Remember that from Stage 1's perspective, he is stationary in space and the Earth is moving away from him in the other direction at 37,701 miles per second. The speed of light must have a reference; this reference will always be the observer. If there are two or more observers moving at different velocities in reference to each other and the same light source, the speed of light they detect will always be "c" to each of them.

In 1904 Henri Poincaré wrote:

"From all these results, if they were to be confirmed, would issue a wholly new mechanics which would be characterized above all by this fact, that there could be no velocity greater than that of light, any more than there is a temperature below that of absolute zero. For an observer, participating himself in a motion of translation of which he has no suspicion, no apparent velocity could surpass that of light, and this would be a contradiction, unless one recalls the fact that this observer does not use the same sort of timepiece as that used by a stationary observer, but rather a watch giving the "local time"

(Poincaré, 1904, pg. 253)

64

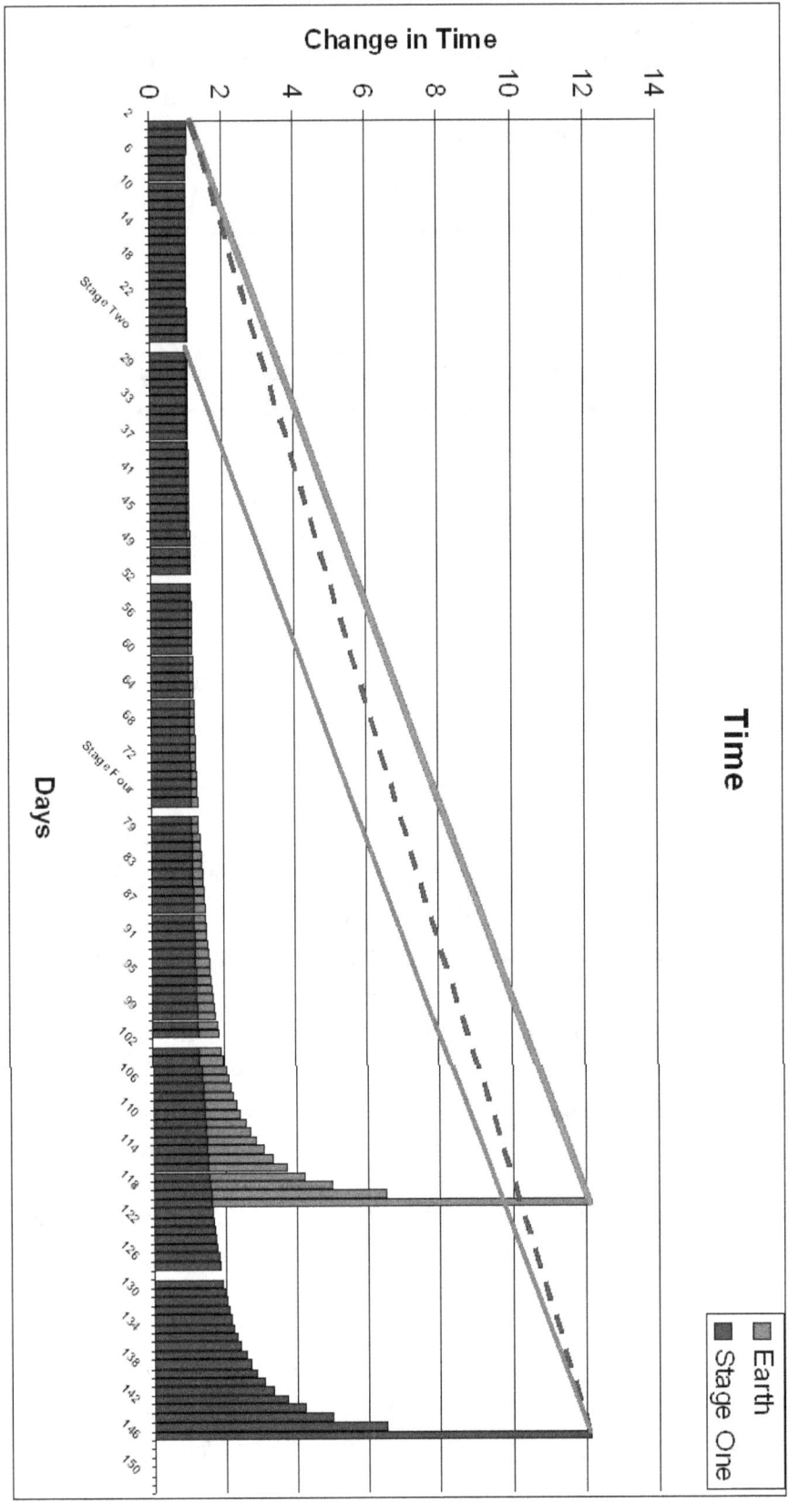

CHAPTER IV

RESEARCH METHODOLOGY

Research Design

Most of the data has been drawn upon from the theories and ideas in the writings of the architects of Quantum Mechanics such as Albert Einstein, Max Planck, François Arago, Max Born and Stephen Hawking whose ideas had root in the teachings of philosophers Aristotle, Copernicus, Galileo and Sir Isaac Newton of centuries past.

Research Model

The research model is based on the history and evolution of natural science into the study of physics and modern day quantum mechanics. The fundamentals are laid out, researched and explained, most with thought experiments that think through or walk through the process to ensure it is rational and then worked out mathematically. Each step in the evolution process is explained so that the layman without a mathematical or scientific background can understand the underlying fundamental concepts of General Relativity the Quantum Field Theory.

General Relativity relates to Gravity in relation to the macro and the Universe at large around us. The Quantum Field Theory is engaged in the micro, the subatomic universe and the three non-gravitational forces referred to as the strong, the weak and electromagnetic forces which regulate and balance the subatomic world. Einstein would try to unify these two physical theories in his later life and even on his deathbed struggled to create a Grand Unified Theory where the laws of the Macro Universe and the laws of the Micro Universe work hand in hand. Einstein would also say, *"Who would have imagined that this simple law has plunged the conscientiously thoughtful physicist into the greatest of intellectual difficulties?"* (Einstein, 1952, pg. 18)

This study addresses many of these conflicting theories and suggests possible alternative theories that allow these two worlds to work as one. These ideas, many of which have been researched and discarded many years ago are readdressed as still possibilities if understood in a different light. The primary example that the Aether Theory was openly rejected after the Michaelson Morley Experiment of 1887, yet light is clearly manifested as transverse waves that can only propagate through solid mediums. If this medium does not exist, how can a wave be a wave? What is waving? If something consists of a field, it is a field of what? Time is also referred as "local" as various systems in the Universe have time manifesting its self at different rates due to gravity and acceleration forces, seemly indicating that space varies in density. Density suggests that the attributes of "empty space" are denser or less dense and that even empty space consists as a field of something. Energy transfers through a substance in this space at different rates, energy manifesting its self as the movement of mass.

Sources of Data

The sources of the information in this research and proposal come from some of the following and much more which are listed in the references at the end of this research paper.

Albert Einstein, *Relativity by Einstein* and *Special and General Theory*

François Arago, *Oeuvres Complètes*

Aristotle, *De Caelo*

Isaac Asimov, *Atom, Journey Across the Subatomic Cosmos*

Marcia Bartusiak, *Einstein's Unfinished Symphony*

Max Born, *Einstein's Theory of Relativity*

Galilei Galileo, *Dialogue Concerning the Two Chief World Systems*

Stephen Hawking, *A Brief History of Time* and *Hawking on the Big Bang and Black Holes*

H.A. Lorentz, *The Theory of Electrons*

Sir Isaac Newton, Opticks, *A Treatise of the Reflections, Refractions, Inflections and Colors of Light*

Most of the study is also compiled in two of these researchers' books.

Mark F. Dennis, *The Speed of Light, Reviewing the history of "c"* and *Traveling at the Speed of Light*

CHAPTER V

QUESTIONS AND THEORIES

Back to Basics

In the late 1890's, the pace of scientific discovery was colossal. Through the discovery of x-rays and other types of radiation, scientists such as Henri Becquerel as well as Pierre and Marie Curie began to put together the atomic model. Just after the turn of the century in 1911, Ernest Rutherford concluded through his research that a positive charge was confined to the small region at the center of the atom, what he called the nucleus; this was only fifteen years after the discovery of nuclear radiation by Becquerel and the Curie's (Bord, pg. 441). Several years later, in 1913, Danish physicist Niels Bohr published an atomic model that better explained the natural atomic spectra using the now well-known "solar system" as a nucleus and electrons in distinct orbits around the atom (Bord, pg. 381). The original model only subscribed to a positive and a negative charged particle model until in 1935, James Chadwick, working with Irene Curie's work (Marie Curie's daughter) discovered the neutron.

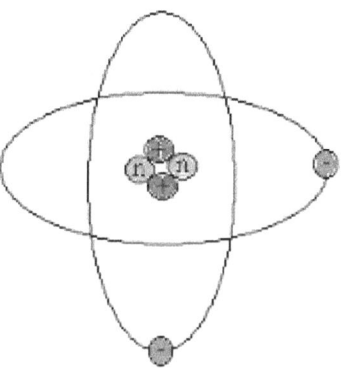

Bohr's Model of the Atom

The electron is the life of the atom. Electrons are believed to travel around the nucleus of the atom at fantastic speeds, some 700 miles per second and higher. Scientifically speaking, the electron is assigned a negative charge and the proton a positive charge, with opposite charges theoretically attracting each other equally. It is this high velocity in orbit that keeps the electron from falling into the nucleus of the atom as a result of the tremendous mutual attraction of the electrons and protons charges.

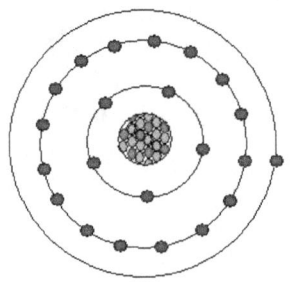

Conductor - One Electron in Outer Shell

Although the forces of the electrons charges are equal to that of the proton, the electron is approximately one trillionth of an inch in size with the proton around 1,800 times the electrons size. The neutron is similar in size to the proton but has no charge. A stable atom of any element normally carries a balance of multiples of these three atomic particles. A positively charged atom will carry more protons than electrons (a deficiency in electrons) and a negatively charged atom will carry more electrons than protons (Veatch, pg. 22-23).

Today, about 155 combinations of these particles are known – called elements. Elements with a different number of neutrons than protons may have the same characteristics chemically but are considered an isotope and may be radioactive, behaving differently from a nuclear perspective (Bord, pg. 413).

Some elements such as copper are called conductors and have the ability to share electrons in the outer orbits (shells) of their atoms. As a rule of thumb, elements with one, two or three electrons in their outer orbit are considered conductors, such as gold, silver and aluminum. These electrons move quite easily from atom to atom; this flow of electrons is called electricity. This sharing of electrons also produces the Covalent Bond that binds atoms together creating various chemical structures.

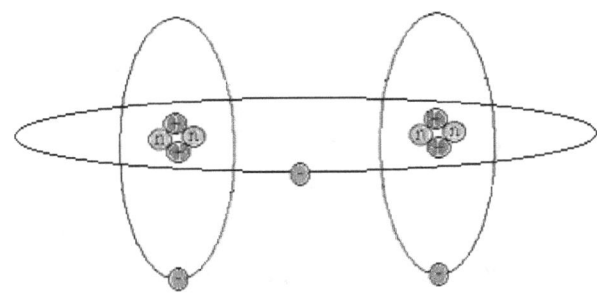

Conductor Sharing an Electron

Other elements have their outer orbits completely full and do not share electrons. These elements are considered insulators and resist electrical currents because their valence shell (outer shell) is full with valence electrons. The more electrons in these outer shells, the more inert the element is. Materials that fall in between insulators and conductors are called semiconductors such as carbon, silicon and germanium and are used primarily in electronic circuitry and integrated circuit chips (IC) used in electronics today (Evans, pg. 2).

Whether in chemistry or electronics, the electron is a source of energy that can be tapped. Electricity can be used to produce heat, magnetic energy in motors and electrochemical reactions such as chrome plating. Electricity is also used to create light and various types of electromagnetic propagation such as radio waves, microwaves, x-rays and gamma radiation. It is in understanding the electron that we can start to understand electromagnetic phenomena, energy and the fundamentals of light.

Energy

What is energy? Many eastern religions and even the new age movement speak of pure energy. Is there such a thing as "Pure Energy"?

Scientifically speaking, energy is a variable quantity that is a characteristic of an object. Energy is the ability to do work or that which is transferred when work is being done (Bord, pg. G-1). There are several ways to describe energy, kinetic, potential, gravitational, electromagnetic, chemical and nuclear. Energy can be transformed from one form of energy to another such as light into heat, but cannot be created nor destroyed. This is known as the Law of the Conservation of Energy (Bord, pg. 98) written in Sir Isaac Newton's *"Philosophiae Naturalis Principia Mathematica,"* Latin for "Mathematical Principles of Natural Philosophy"- one of the most influential scientific works ever published.

Let's start with some energy formulas:

Velocity = Distance/Time - the rate of change in the position of an object over time.

Acceleration = Velocity/Seconds^2 Change in velocity of an object over time, over time.

Momentum = Mass x Velocity - the mass of an object multiplied by its velocity.

Inertia = A resistance of an object to the change in its state of motion.

As we can see, all of these formulas contain an object. Objects have mass, therefore "energy" simply put is a quantity associated with the movement of mass either kinetically (active) or potentially.

As a ball is thrown, it is accelerated to a specific velocity and has a certain momentum; the ball then decelerates and falls to the earth. The energy put into the ball dissipates through friction in the air and gravity. The remaining energy the ball has then is transmitted into and dissipated through its contact with the ground as vibrations and as sound through the air.

A rocket shot into space will remain at the velocity it was accelerated to once the engines are shut down. It will stay at this velocity and direction of travel until acted upon by another force such as gravity or friction due to reentry into the atmosphere.

Newton's First Law of Motion

"Every object in a state of uniform motion tends to remain in that state of motion unless an external force is applied to it"

(Bord, pg.47)

Einstein's Most Famous Equation

As we look at Einstein's most famous equation we also see that mass has to be part of the energy equation.

$$E = MC^2$$

Here E is energy, M is mass and C is the speed of light at approximately 300,000 kilometers per second or 186,000 miles/second - identical to momentum - Mass x Velocity (Born, pg. 283-286).

Subatomic Particles

There is a difference in how subatomic or elementary particles react with other particles and how they dissipate their energy. Unlike our world, where energy such as sound is dissipated through many different *medium*s and distributed into billions of particles, energy from subatomic particles can only be distributed and dispersed into fixed packets or bundles of energy that cannot be divided (Polkinghorne pg. 6).

In the twentieth century, studies revealed that light could cause atoms to emit electrons. In 1900, while studying Black Body Radiation (BBR), Max Planck (1853-1947) found that atoms emitted energy in multiples of a certain energy unit, E or 1, 2, 3… and so on, multiples of E; it was not possible to emit a fraction of a unit (Polkinghorne pg. 5). Einstein suggested that light energy may not be evenly spread across a given area but concentrated in small "packets" or "quanta". The energy transferred from one particle to another had to be completely transferred, all or nothing.

This localized energy was coined the "photon" (Kafatos, pg. 29). The "aether theory" was thrown out and the "corpuscle theory" also known as the "particle theory of light," originally suggested by Newton, was then reintroduced. The "aether theory" had too many questions that could not be answered. The "particle theory" solved many of these questions and then became the preferred theory of the scientific community.

<u>Subatomic Expressions of Energy</u>

Subatomic particles such as electrons demonstrate the energy they obtain in several ways.

Spin – Where the particle has angular momentum, a conserved quantity based on its mass and speed that will continue unless acted upon by another force (Polkinghorne pg. 96). The spins of all known particles consist of a half spin or whole spin or a multiple of that factor, for example 0, ½, 1, 3/2 … etc. This energy is transferred from one particle to another in fixed quantities in particle-like packets of energy called quanta (Bord, pg. 458).

Velocity – Where a particle has orbital angular momentum in orbit around an atom. This energy is also transferred from one particle to another in fixed quantities in particle-like packets of energy called quanta. As an electron absorbs a specific "quanta" of energy, its velocity and energy level change and cause it to "jump" from a lower orbit to a higher orbit. Likewise, as an electron emits or loses a "quanta" of energy, its velocity and energy level again changes causing it to "jump" from a high orbit to a lower orbit (Bord, pg. 381).

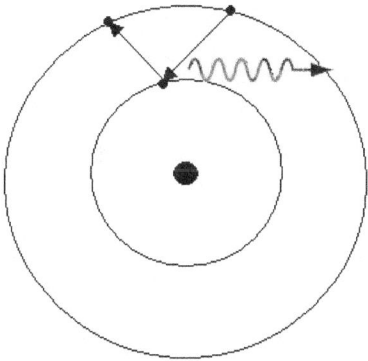

Electron Releasing a Packet of Energy

Frequency – The rate at which a particle oscillates per second (Polkinghorne pg. 213). This frequency is proportional to the wavelength of the energy it disperses. This wavelength is the measured distance of a cycle from beginning to end. This electromagnetic energy wave travels at the speed of light (29,979,245,800 centimeters per second) and their frequency and wavelength can be determined by these formulas:

Frequency = <u>Speed of light</u>
        Wavelength

Wavelength = $\underline{\text{Speed of light}}$
                Frequency

Speed of light = λ x Frequency   Wavelength = λ

The amount of energy a particle such as a photon has depends on the frequency at which it resonates. $E = hf$ where E is energy, h is Planck's constant at 4.136×10^{-15} Electron Volts (eV) and f is the frequency that the particle oscillates at in Hertz (Bord, pg.375).

Examples:

Red light: $f = 4.3 \times 10^{14}$ HZ (Hertz) at 1.78 eV

Blue light: $f = 6.3 \times 10^{14}$ HZ at 2.61 eV

X –ray: $f = 5 \times 10^{18}$ HZ at 20,700 eV

An electron in orbit is balanced between electrical attractive forces that pull it towards the nucleus while its centripetal acceleration caused by angular momentum keeps the electron from falling towards the nucleus. In the lowest orbit, orbit one; an electron has the smallest amount of energy possible. As it gains energy the electron moves out to larger orbits. Each "jump" to another orbit is caused by a gain or loss of packets of energy called "quanta" – Planck's constant. Each jump called a "radiative transition" causes the electron to radiate or absorb energy and change orbit. The larger the transition or "jump", the higher the frequency of the energy radiated. For example, the transition from orbit 3 to orbit 2 may give off the lower frequency red light burst whereas a "jump" from orbit 6 to 2 may give off a high frequency violet light burst. Electromagnetic spectroscopy analysis of these light burst show patterns that are unique to specific elements and can be used as fingerprints in determining the element present by the light they radiate.

The speed of an electron, as calculated by French physicist Louis de Broglie (1892-1987), in the smallest orbit of a hydrogen atom is about $2.19*10^6$ m/s or meters per second (Asimov, pg. 86). The speed of light is $2.998*10^8$ m/s or 299,800 kilometers/sec (186,000 miles/sec). The mass of an electron is approximately $9.11*10^{-31}$ Kg. Therefore the momentum of an electron in the smallest orbit at $2.19*10^6$ m/s is $1.995*10^{-24}$ kilogram-meters per second (Bord, pg. 385).

If this momentum is transitioned to the speed of light though the photon, the mass of the photon would be the energy or momentum of the electron divided by the velocity of the photon which is the speed of light or $(1.995*10^{-24})/(2.9979*10^8) = 6.655*10^{-33}$ Kg (compared to the electrons $9.11*10^{-31}$ Kg). This seems to suggest that the photon would be 136.89 times smaller in mass than an electron and 251,532 smaller than the mass of a proton.

Likewise, if a photon's mass is 6.65496 *10^{-33} Kg. and its velocity is the speed of light at 2.998*10^8 m/s then = 1.995*10^{-24} kilogram-meters per second – enough momentum to cause an electron to "jump" from one shell to another in an orbit around an atom or the effect of "h", a Planck constant or "quanta".

The equation E = MC^2 indicates that the energy of a photon depends on its mass multiplied by its velocity (the speed of light) or E (photon) = mv^2. In reference to the speed of light this would be E = c (mv) or E = cP where (mv) or P is momentum.

There is a problem though; according to Einstein, as a particle increases in velocity closer to the speed of light, its mass increases exponentially – called relativistic mass compared to mass at rest. For example if an object whose mass is equal to 500 kg approaches the speed of light, say 90% the speed of light, its mass would increase to approximately 1143.8 kg - more than double its mass.

$$m = \frac{m_0}{\sqrt{1 - \dfrac{v^2}{c^2}}}$$

Relativistic Mass Formula

According to physicist, Satyendra Nath Bose, the photon must be different from other subatomic particles in that it must have zero mass at rest in order to go the speed of light (Isaacson, pg. 328). As the velocity (here referenced at c, the speed of light) of a particle decreases, the mass of the particle also must decrease to the point where it has no mass at rest (Ditchburn, pg. 343).

All these fifty years of conscious brooding have brought me no nearer to the answer to the question, 'What are light quanta?'

Albert Einstein   (Isaacson, pg.101)

74

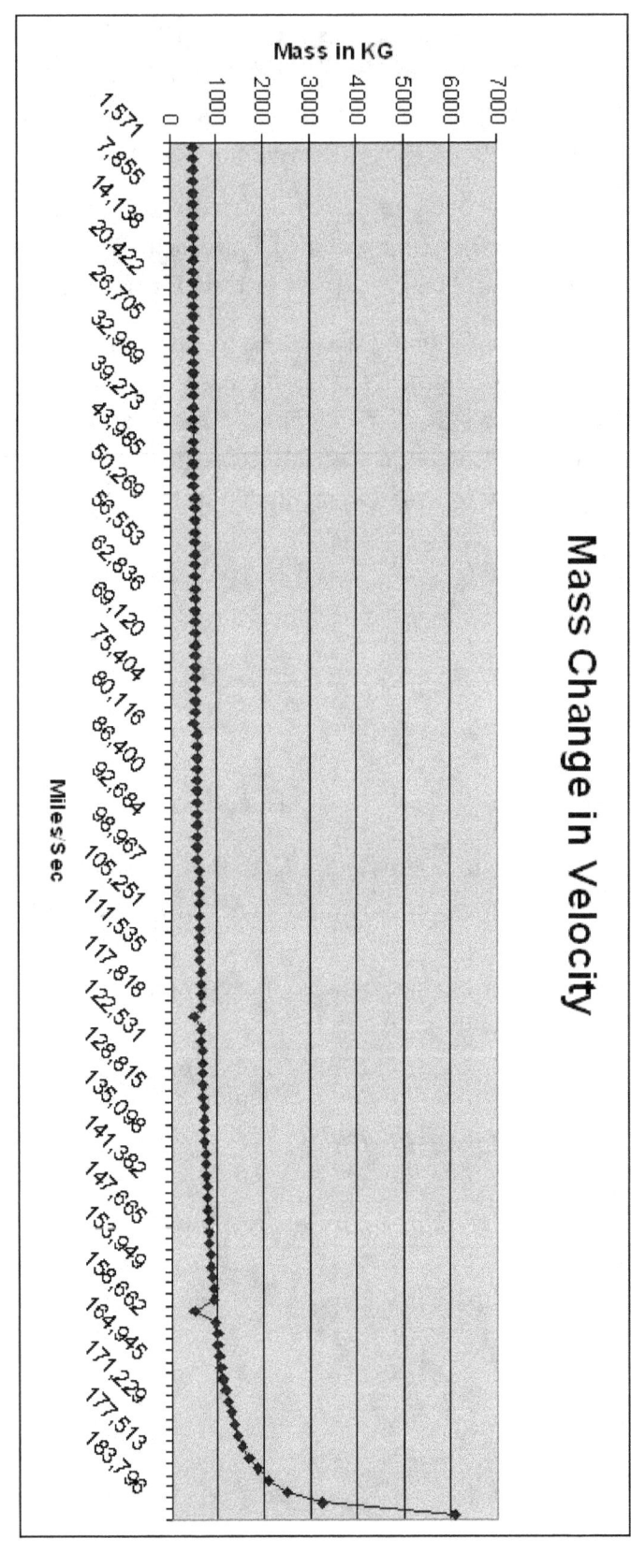

75

Quantum Questions

The following examples will not use the Lorentz Transformations whose numbers exponentially compound, but rather standard multiplication and division for simplicity sake, however the principle is the same.

Let's go back to the first illustration of the Theory of Relativity with the Earth moving towards and away from a star in the beginning of this book.

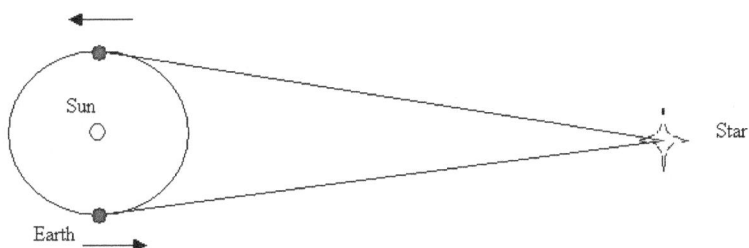

Movement of the Earth

Earth toward a Star:

186,000 m/sec +18 m/sec = 186,018 m/sec

Earth away from Star:

186,000 m/sec - 18 m/sec = 185,982 m/sec

Our dilemma here then is that:

186,000 m/sec +18 m/sec = 186,000 m/sec

And

186,000 m/sec - 18 m/sec = 186,000 m/sec

One approach to change this is to change time:

Lengthening the second or Time Dilation

1 second = 1.00009674 <u>seconds</u>

Where 186,018/ 1.00009674 = 186,000m/sec

Another approach to change this is to change space:

Shortening the mile or Length Contraction

1 mile = .999032 <u>miles</u>

Where <u>186,018</u> miles X .999032 = 186,000 miles

Note that both end results are 186,000 m/sec no matter what increase or decrease in the detectors or the emitter's velocity. Also note that any combination of space and time values that result in the same value is also possible.

Analysis:

Now let's view this with a spaceship. As the spaceship moves towards the star at a high rate of speed, in order to compensate for the speed of light coming from that star, time and space must change. The spaceship at this velocity **contracts** making everything on spaceship in the plane moving towards the star shorter. Time also is **dilated** allowing for the speed of light from the star to remain at "c". Also from the time perspective, to the stationary observer in space, a clock on the spaceship must then tick slower.

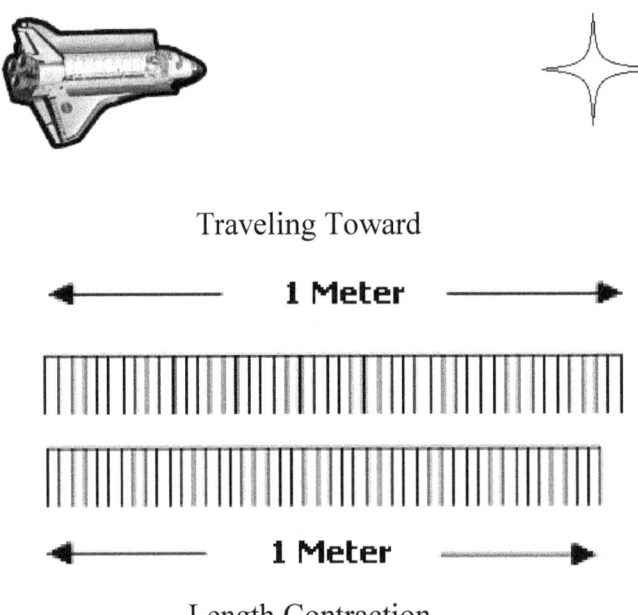

Traveling Toward

Length Contraction

Now, let's move in the opposite direction. As the spaceship moves away from the star, the spaceship theoretically must stretch in the plane moving away from the star to compensate for the speed of light. Time must also then be must compressed to compensate for the constant speed of light. From the time perspective, to the stationary observer in space, a clock on the spaceship must then tick faster.

Traveling Away

1 Meter

1 Meter

Length Dilation

In another thought experiment, let's put two spaceships with light measuring devices onboard moving in opposite directions relative to a light source (star). These two devices will travel at a high rate of speed in opposite directions reading the stars light. We will observe these two devices from a stationary observer's viewpoint.

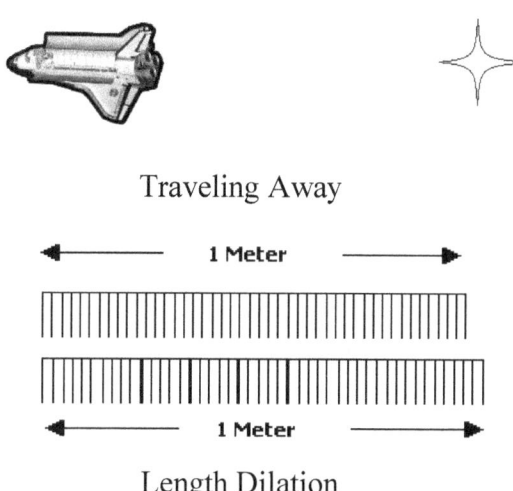

Observing Two Moving Objects

Device No.1 is moving towards the star and device No.2 moving away from the star are both reading the same light from the star. Device No. 1's length will contract to compensate for the speed of light and its clock ticks slower than the stationary observer. Device No. 2's length will have to be stretched and its clock will have to tick faster in order to compensate for the speed of light in reference to the stationary observer. These two devices will in fact be emulating the Earths motions towards and away from a star (at different times of the year) at the same time as in our first example.

Now, let's look at it from a different perspective. We will now only have one moving measuring device observing the light from two stars.

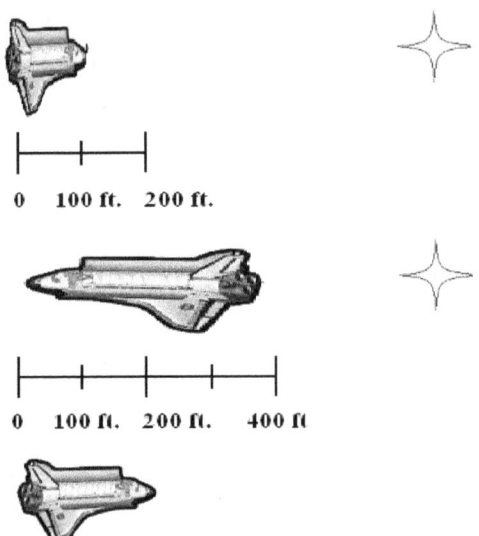

Observer 2

To the stationary observer, the clock on the device moving towards star No.1 ticks slower and its length contracted to compensate for the speed of light emitted from star No.1. However, the stationary observer also must perceive the clock on the measuring device moving away from star No.2 as ticking faster and its length stretching to compensate for the speed of light from that star. How can this be? From the same observer, the same object is both speeding up and slowing down and at the same time contracting and stretching to compensate for the speed of light. This appears to be a paradox.

0   100 ft.   200 ft.

0   100 ft.   200 ft.   400 ft

Length Contraction and Dilation

A Challenge to Einstein

At first glance one might want to challenge the Michelson Morley Experiment by pointing out that the experiment would not work. It would not work simply for the fact that one beam crosses the aether wind at 90^0 and the other beam flows directly into the aether wind. It is being slowed down at first, but after being reflected off the mirror is sped up while flowing with the aether for a null effect.

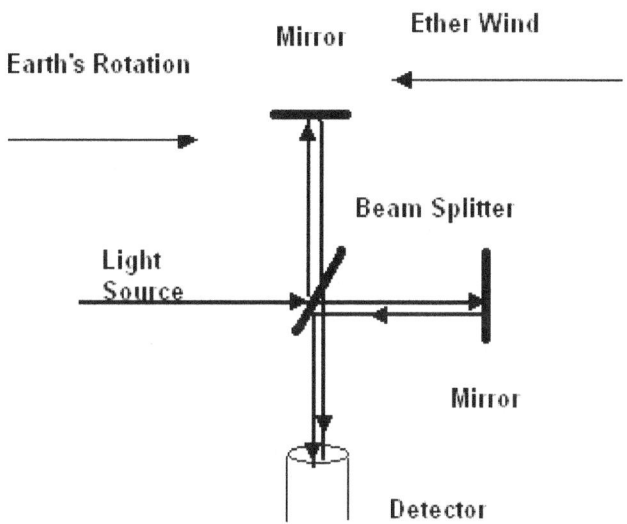

The Michelson Morley Experiment

Let us use the flights of some aircraft for an example to emulate this experiment and compare two flight plans, both 2,500 miles one way. For the first aircraft, let's assume that a flight from Miami, Florida to Goose Bay, Newfoundland, Canada is both 2,500 miles one way and directly north in heading. For the second aircraft, the flight is also 2,500 miles from Boston, Maine to Seattle, Washington and is directly west in heading. The wind is straight out of the west at 100 knots at altitude. Both aircraft will maintain 500 knots airspeed the whole flight at the same altitude.

Using our aircraft navigation computers we take off with the two aircraft at the same time and fly round trip from Miami to Goose Bay and back and from Boston to Seattle and back, at the same air speed and wind conditions, timing both aircrafts round trip flights.

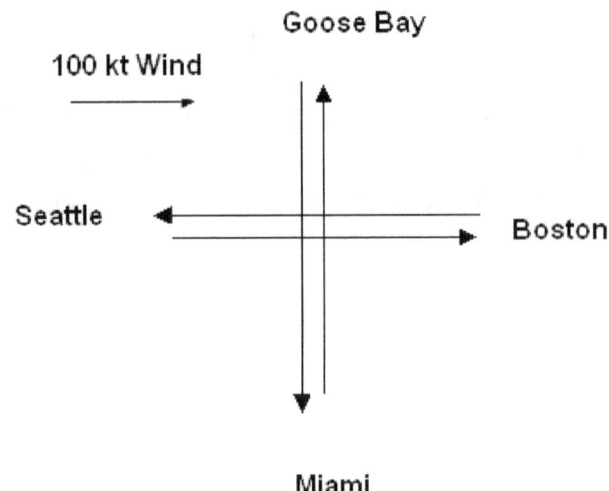

Flight Path

The aircraft flying to Goose Bay averages 510 knots ground speed going to Goose Bay and 510 knots ground speed back to Miami with a 90^0 crosswind both ways. It took 9.804 hours round trip to complete the flight. The second aircraft had to first fly into the wind and then return with the wind. Its first leg to Seattle took 6.25 hours while it fought the wind; the second flight back to Boston, however, only took 4.166 hours flying with the wind. The difference might surprise you; round trip for this second flight was 10.4166 hours, 37 minutes longer than the north south flight. Why the difference? Both flights had to fight the same wind and both flights were 5,000 miles round trip at the same airspeed! The reason? Diminishing returns! The harder the wind, the more time lost. At one point as you raise the bar, the wind will be at a higher velocity than the aircraft can fly and the aircraft will in fact fly backwards in reference to ground speed.

How does this relate to the Michelson Morley Experiment? The Michelson Morley Experiment was an experiment whose intent was to discover the "aether wind." As we can see in our aircraft experiment, there was a difference, but what if the wind was only one or two knots instead of one hundred knots? Would we have noticed the difference? How fast is the "aether wind"? Would the wind be one mile per hour? Would it be five thousand miles per hour? If this experiment would measure anything, how about pointing it up and down? Would it measure the effect of gravity on the light in the vertical plane? Isn't gravity, after all the "aether wind" anyway?

The "Big Bang"

Hubble's discovery in 1929 of an expanding universe lead scientist to try to determine what happed in the past by looking at the Universe backwards in time, some 10 to 20 billion years ago to a space-time "singularity" where everything was "one". All matter would come from a certain point where measurements were infinite, as if in a black hole (Asimov, pg. 305).

This is how it may have started:

In the beginning, there may have been no such thing as time and space; then something happened. An immense release of energy occurred; a critical stage was reached, possibly similar to "critical mass" in a nuclear event out of the "singularity". A point was achieved in which the accumulation of mass and thermonuclear energy could no longer be sustained and a massive sub atomic nuclear like explosion occurred. In this huge explosion, existing mass is blasted into the smallest of particles as a hot plasma gas out into the far reaches of space. Huge amounts of energy are released as these smallest of particle ricochet off each other at enormous velocities, dissipating their energies out into the blackness of the universe.

Over time this energy continues to spread out from the center of the energy burst. As this outward energy dissipates into the regions beyond, a reversal of the flow or an "implosion" of these particles of mass begins to take place, back toward the center of the blast occurs. As time passes and as the particles energies and spins start to subside, higher energies and pressures now exist to the exterior of this spherical blast, pushing these small bits of matter together forming a denser pocket of matter in the low pressure center. Now cooler at the center, where the blast first took place, these "particles" now begin to go through a phase transition, reaching a *critical point* and precipitate or condense, forming a denser core – the beginnings of subatomic particles and atoms.

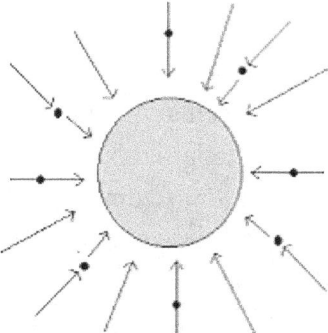

Condensation of Particles

One analogy that can be used is to look at these "particles" as low-pressure systems with pressure gradients such as isobaric lines that slowly decrease in pressure from the outside in. Just as miniature hurricanes or "black holes", material or energy in this material slowly spirals in towards the center of this blob of spinning mass.

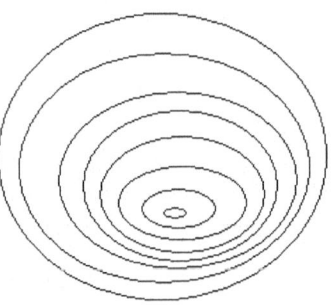

Pressure Gradients

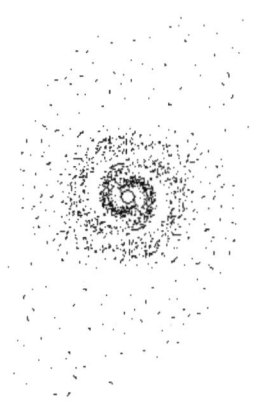

Low Pressure System

Over time, as energy disperses out into space, more and more of these "low pressure systems" begin to form. Elementary or fundamental particles flowing into these systems are spread out over great distances, spiraling in towards the center of these condensing storms. As these different systems come closer together they begin to react to each other even at great distances, as the space between them is not empty. Low-pressure systems spinning in the same direction repel each other even though far apart, bouncing off each other like tops spinning the same way. Systems spinning in opposite directions move together like gears in a clock, attracting one another and moving in unison.

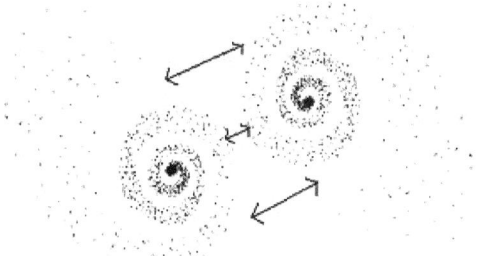

Like Spins Repel

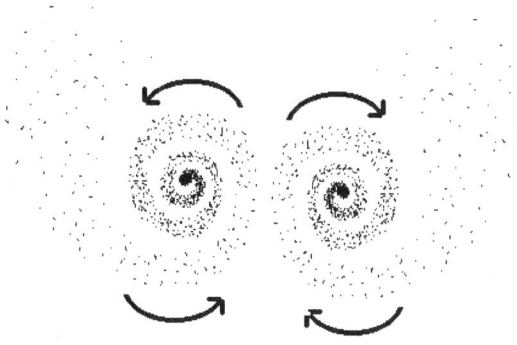

Opposite Spins Attract

As these systems or particles begin to interact with one another they transfer energy, bouncing off of, spinning with and spinning around each other until equilibrium of energy is attained. Various sizes of particles of different spins pair up with particles of opposite spins creating balance. Smaller particles of less mass with spins opposite of larger particles attract each other causing the smaller particles to begin to spin around the large particle, entering an orbit. The orbiting particles become the life of these systems, speeding up and slowing down as they absorb or give off energy. The same number of particles in orbit balances the number of particles in the nucleus being orbited and an atom is formed.

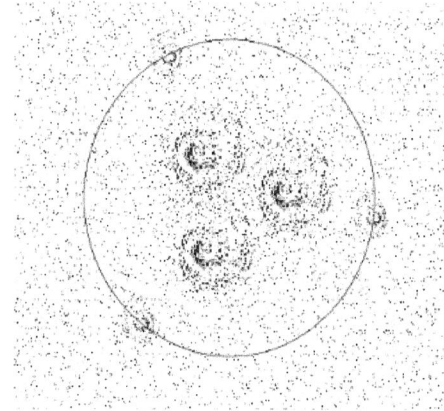

Particles Moving in Unison

The amount of energy given off or absorbed is directly proportional to the size of the particle and its velocity in orbit around the central system or nucleus. This amount of energy is a fixed value, a packet of energy, based on the velocity of the orbiting particle in the lowest possible orbit. This energy however, can be passed on to other particles at multiples of that value. As these orbiting particles absorb energy, they increase in momentum and move further away in their orbit around the nucleus. Also as these particles give off energy, they decrease in orbital momentum and move closer to the nucleus.

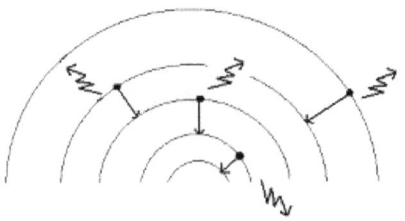

Release of Energy

As energy emitted from these systems is dispersed in these whole value "packets" of energy, they are likewise absorbed as whole packets of energy in other systems, even at some distance. The constant giving off and absorption of this energy makes the particles in orbit to move in and out in their orbits causing the nucleus to oscillate forming a wave pattern or "matter wave". This constant waving in and out of the orbiting particles disturbs the cloud of tiny particles that fills the space between these particles causing other orbiting particles even at great distances to resonate in unison. Particles in different orbits resonate at different wavelengths dissipating energy at various frequencies through the space around them.

Simply stated, the hypothesis stated here is that existing matter is made up of the smallest of particles that are gradually condensing as their energy dissipates out into the universe. These particles become the atoms, protons, neutrons and electrons that create all things in our universe. Energy is simply the movement of these particles in various ways and the interaction of these particles with each other even at great distances. The space between subatomic particles such as electrons and protons is not empty and the movement of one particle will ultimately move other particles transferring energy from one area to another at the speed of light even though far apart.

According to the general theory of relativity, space without aether is unthinkable; for in such space there not only would be no propagation of light but also no possibility of existence for standards of space and time.

Albert Einstein (May 1920)

(Isaacson, pg. 318)

Particles

As we saw previously, particles as we know of as electrons, protons and neutrons are possibly condensates of elementary or fundamental particles. These condensates may be formed by pressure gradients created by the cooling of a gaseous mass made up of these particles. During cooling, these condensates reach a saturation point and transition phase, bonding together at a specific energy level into a unified particle. Similar to the dew point, this may occur when specific pressures and temperatures are reached and "droplets" form. These droplets can only exist as a quantity of these elementary particles at specific energy states. These particles can be in a stable state as condensates of cooled fundamental particles formed into protons and electrons. They also may be manifested as moving corpuscle particles, for example, a photon that is produced by pressure waves created through compression.

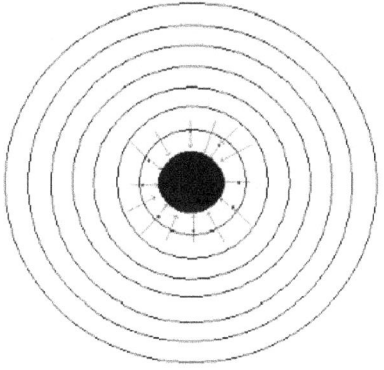

Condensates of Elementary Particles

These corpuscles or discrete particles are formed by pressure waves from energy traveling through an elastic *medium* of elementary particles. Again, a wave is defined as *"a traveling disturbance of coordinated vibrations that transmit energy through a substance without net movement of that substance."* These discrete particles are created by energy passing through these relatively stationary particles.

These corpuscles are created by changes in the density of this *medium* from waves of particle motion in a specific direction. Since these particles can only exist under specific energy conditions, they remain confined to only regions of pressure or pressure gradients that create this phase change. In the form of an electron, this energy of moving waves of particles (spherical standing waves possibly) will be confined to a specific pressure level in the *medium* that exists as rings or shells around the nucleus of an atom. As energy is absorbed or emitted from these pressure gradients, existing electrons will dissipate into the *medium* of the region they are presently in and reform as a condensate of energy waves in particles of a different pressure gradient or shell surrounding the atom – disappearing and reappearing. Electrons may also be viewed as a sort of hybrid, a stable condensate particle yet wave-like in form. Energy given off or absorbed by these electron waves is only transferred in fixed "packets of energy", allowing these energies to only exist in specific shells specified by their momentum; the energy of the smallest shell or orbit possible, contains the energy of one of these "packets of energy". Each shell would then be created by multiples as if harmonics of this energy.

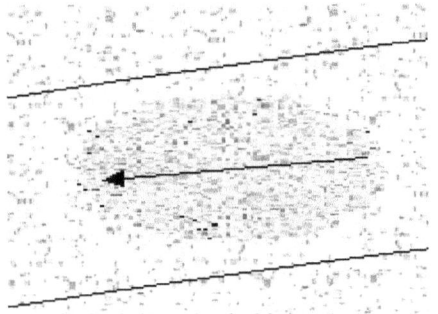

Electron as a Corpuscle of Energy

The energy that is being emitted or absorbed by these regions surrounding the atom is in the form of another discrete particle called the photon. This particle also exists as a pressure wave through the *medium* that can only exist under specific conditions as a "packet of energy". This packet of energy is given off by an electron as a pulse of energy in a given direction. The electron or wave of energy caught in the orbit of an atom that is emitting this energy then dissipates and reemerges as a wave of energy in a lower shell closer to the atom. The wave of energy given off from the electron as a pulse, then moves as a corpuscle of changing density in the *medium* of elementary particles in a straight line at the speed of light. This energy continues in that direction until it is absorbed by another electron in the orbit of another atom far away. Since energy cannot be created or destroyed and this particle can only exist under specific energy levels, energy is concentrated in one direction and cannot be dispersed. No "elementary particle" of this *medium* continues to move in this direction; they simply oscillate in place passing energy from one particle to another, compressing together into a denser mass in a chain reaction in one direction at the speed of light.

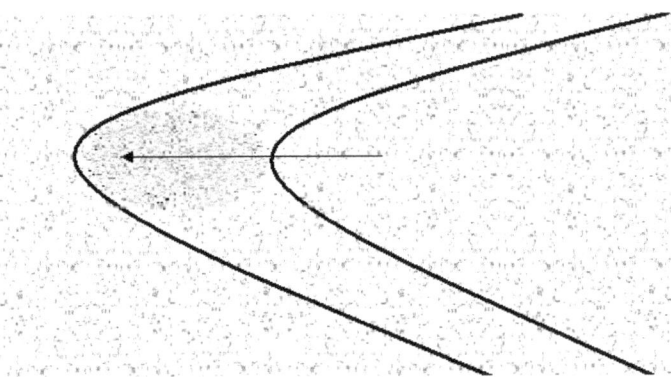

Photon as a Corpuscle of Energy

The photon as a particle, unlike the electron and proton, can only exist as a wave and has no mass of its own, just as a wave in the ocean cannot exist by itself. The photon and the wave on the ocean are simply the manifestation of energy – the movement of mass. The mass of the photon in a snapshot of time is the sum of "elementary particles of the medium" packed together in the wave front and may vary according to the density of the medium and the velocity of the wave through the medium. Its momentum or energy will vary as summations of "packets of energy". This process is continuously generated as electron waves surrounding atoms constantly absorb and emit this energy causing a back and forth waving of these particles in the *medium* of elementary particles. The different intervals or timing of these cycles depends on the energies and the type of atoms giving off this energy, producing different wavelengths of photon waves or colors of light detected in our eyes.

Electrons and photons do not only travel at high velocities in a given direction or orbit; they also oscillate back and forth at a given rate as they move in this direction. The energy of these electrons and photons can be mathematically represented in wavelengths. The lowest orbit possible would hold one packet of energy known as the Planck Constant and one full oscillation of the electron in its orbit would be explained as one full wavelength. An electron in higher orbits will have a higher a higher angular momentum, therefore a higher frequency of its oscillations and a shorter wavelength. The formula $p = h/\lambda$ can be used to show the momentum or energy of electrons in each orbit. Here p is the momentum, h is the multiple of Planck's constant in whole numbers and λ, the Greek letter lambda, and is the wavelength of the electron (Kafatos, pg. 33). The formula expressed as $\lambda = h/p$ shows the wavelength of the electron based on the multiple of the constant divided by the momentum of the electron (mass multiplied by velocity, mv). Energy released by the electron changes its momentum, thus its wavelength. This energy is transferred in another waveform out of the atom in the appearance of a photon of a specific wavelength and energy level.

Momentum or "p" of an electron in the <u>lowest possible orbit</u> is equal to the mass of an electron of $9.11*10^{-31}$ Kg multiplied by the velocity of the electron in this orbit of $2.19*10^6$ m/s (meters per second) which is calculated to be about $1.995*10^{-24}$ kilogram-meters per second. The wavelength or λ, can now be calculated with the energy of one Planck constant (h) to be 6.63×10^{-34} J-s (joule seconds). Since $\lambda = h/p$, then:

$\lambda = 6.63 \times 10^{-34}$ J-s/$1.995*10^{-24}$ kilogram-meters per sec.

or

$\lambda = 3.32 \times 10^{-10}$ meters or .332 nm (nanometers) the de Broglie wavelength discovered by Louis de Broglie in 1923 also known as the Matter Wave (Bord, pg. 385).

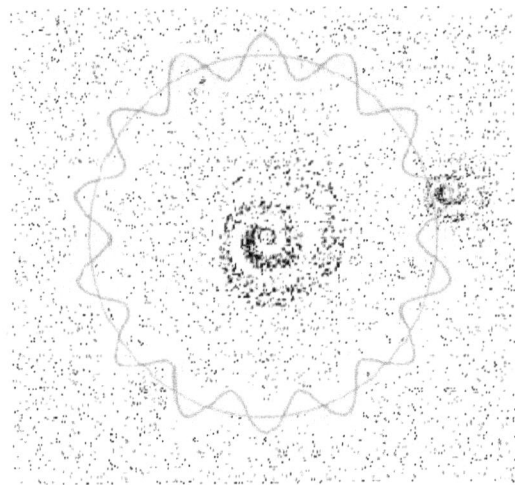

The Matter Wave

Particles whether they be waves of particles or particles as condensates are still created out of this medium and this medium is the foundation of mass and is mass. As particles approach each other, pressure gradients between the particles decrease, causing the differential in pressures around these particles, creating an invisible field that draws them together. Therefore photons and electrons will be affected by gravity and other types of fields created by these energies.

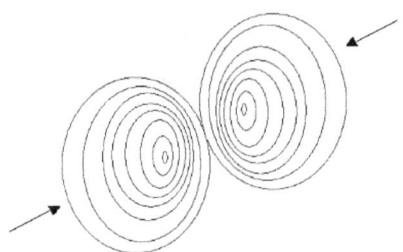

Attraction of Particles

Although all particles may be drawn together in this matter, as seen previously, particles with like spins repel and particles of opposite spins attract. Historically, three subatomic particles are explained in modern physics, electrons, protons and neutrons, in reality only two can exist on their own as stable particles, the proton and the electron since the neutron isolated alone will break down into a proton and electron within 15 minutes.

Studies have shown the masses of the particles to be:

Proton $=1.6726*10^{-27}$
Neutron $=1.6749*10^{-27}$
Electron $= 0.00091*10^{-27}$

What is interesting is that the mass of the electron and the proton added together is = 1.6735*10^{-27}, very close to the mass of the neutron = 1.6749*10^{-27}, a difference of 0.0014*10^{-27}, which ends up to be an additional particle or particles with 1.53 times the size of an electron.

Under extreme conditions during the creation of our universe, high energy levels and atomic pressures turned excess protons in a nucleus into neutrons through the capture of electrons. These combinations of opposite spin particles, not unlike the hydrogen atom, created a neutral particle in the nucleus of the atom helping balance the strong repelling energies of proton, which have same spins and charges.

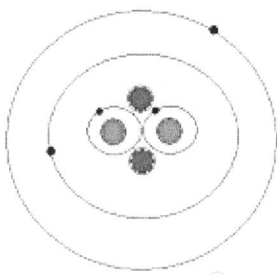

Neutrons in the Nucleus

In summary, a particle can also be viewed as a shockwave that is manifested as a larger single entity created by energy moving through a medium made of fundamental particles. The density and the energy (temperature) of these particles determine the velocity of this shockwave. This wave, similar to sound waves, can only exist at one velocity and cannot go any faster or slower in the material it is passing through. As the density or the temperature of this material changes, so will the shockwaves velocity.

The Photon

Light and other forms of electromagnetic radiation are believed by the modern scientific community to be the bombardment of particles called "photons" or what Einstein called "light quanta". These are particles that are believed to be ejected from the orbits of electrons, moving in a straight line and traveling at the speed of light until being absorbed by another object. Photons, having almost no mass, still have a noticeable displacement of material upon impact raising some questions. Einstein states that the faster a particle moves that it's mass increase until at the speed of light it has infinite mass. How then can anything travel the speed of light and have mass?

In the "Double Slit" experiment performed by Italian mathematician and physicist Francesco Maria Grimaldi (1816- 1863), a phenomenon called optical interference was created which can only occur if light is moving in waves (Ditchburn, pg. 119). In this experiment (also performed by Thomas Young), light from a single source is shone through a single slit in a barrier then continues through a second barrier with Double Slits. Light waves bend or diffract (breaking into pieces in Latin) as they pass through the slits causing the light waves to curve into what is called spherical waves. As light from both slits hits a viewing screen, the wave's peaks and valleys converge at the same points on the screen creating dark areas and light areas of contrast called an interference pattern. Waves that are in phase create constructive interference and brighten the area in which they strike. Waves out of phase cause destructive interference and cancel each other out causing the darker areas on the screen. This optical interference can only occur with light that is correlated or coherent, coming from the same light source (Ditchburn, pg. 116). Monochromatic light, meaning "one color," is more suitable for this type of experiment as only one wavelength is produced creating a more vivid interference pattern; white light is a mixture of all wavelengths and produces a less visible pattern. Similar interference patterns can be also seen with the interferometer that has been adjusted so that the two light paths are of different lengths causing them to arrive at the target out of phase.

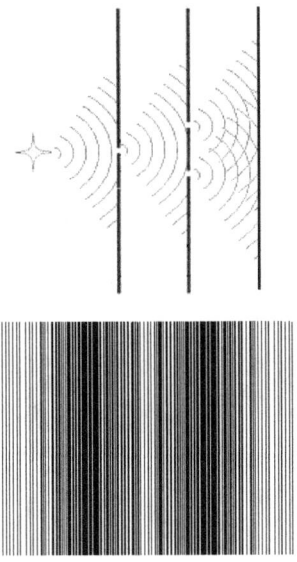

Double Slit Experiment

What causes these waves to diffract as they pass through the slits? This bending of light waves evidently is due to refraction from an interaction with the barrier as the light passes through the slits. The closer the light waves come to the barrier material, the sharper the bend.

The bending of light waves normally occurs when light moves from one medium into another and changes velocity, for example, light transitioning from air into glass. Light travels through air at approximately 185,226 miles per second, or 99% the speed of light in a vacuum. Light passing through glass however, only travels about 114, 989 miles per second, or 62% the speed of light. As the light transitions from air into glass, the light waves slow down and the wave front changes direction causing the light waves to bend or refract.

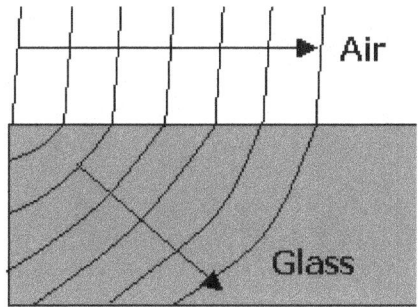

Refraction in Glass

This refraction can be seen with a spoon in a glass of water; light transitioning from water into air after reflecting off the spoon speeds up and changes direction creating the illusion of a bent spoon. The refraction index varies from material to material based on the Optical Density of the material. The denser the optical density, the more the light bends as it passes from one material into another. Optical density is not the same as mass density, however there is a correlation but not linear.

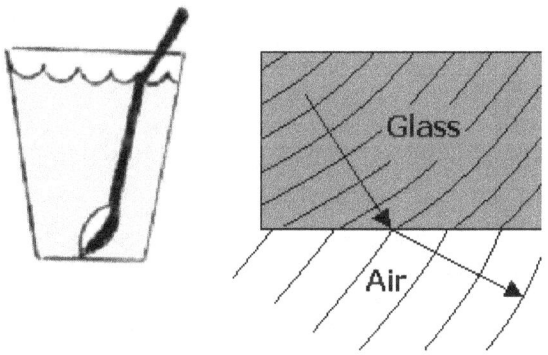

Refraction in Water

The light in the Double Slit experiment bends more the closer it is to the barrier material, as if passing through an invisible field of higher density material. As light passes through this area of higher density it slows down and changes its direction, bending outward. The combined wave becomes curved as a spherical wave; the two spherical waves created by the two slits create the interference patterns on the target surface.

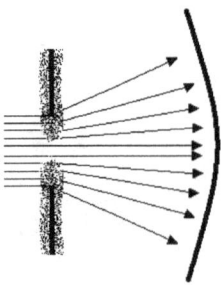

Slit Experiment Diffraction

In 1909 physicist Geoffrey Taylor performed a similar test using dark filters to block most of the photons, allowing only one photon at a time to pass through the slits. Over time the patterns of interference reappeared, showing that single photons will align themselves in a pattern of crests and troughs indicating that the single photon travels as a wave. Electrons also have been fired one at a time creating a similar pattern, refracting as it passes through the slit and falling on the target in crest and trough patterns (Ditchburn, pg. 643).

One at a Time

It is believed that the only way this could happen with single particles is if each photon or electron is divided and passes through both slits at the same time (Hawking, 1988, pg. 61). How else can crests and troughs be created? These patterns rely on interference, the interaction between two waves simultaneously, that create the constructive or destructive patterns on the target, light and dark spots. Otherwise an individual electron or photon would have to exist and then not exist; an on then off switching as it moves through space in order to produce this pattern by its self. For example, when the photon strikes the target when "turned off", it would create a light spot and when striking the target when "turned on", it would create a dark spot. Over time these will form interference patterns on the target. These light and dark spots stay in the same areas, and although created by moving particles, create a standing wave where the visible wave seems to be stationary.

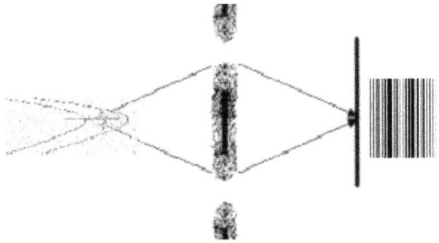

One Photon Two Slits

Let's assume again that a medium exists as if a field around all particles and that this medium moves into these particles through a cooling or condensation process creating varying densities in the medium. Matter attracts matter; therefore, the medium closest to a large mass such as the sun will be denser than the medium out in empty space. As light energy is passed from one particle of the medium to another in waves of various wavelengths, it refracts slightly as it transitions through denser medium, changing the direction it is traveling. Further away in the less dense medium, light waves travel in a relatively straight line. In the Double Slit experiment, light waves passing closest to the walls of the slit curved as they passed through slightly denser medium creating a spherical wave moving towards the experiments target. These spherical waves from both slits created the interference on the target.

Einstein saw a similar event as light waves bend as they pass through the suns strong gravitational field changing the apparent position of a star. This proved to him that light was composed of matter – a particle that is now called the photon. Was this gravity that bent the light waves or the change in the aether's density close to the sun? Possibly they are the same thing.

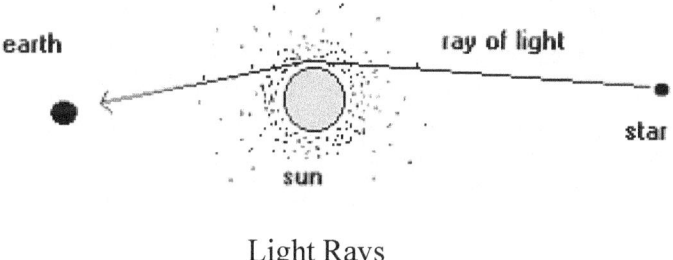

Light Rays

The flow of electrons also affects this medium that surrounds all things, generating electromagnetic waves in various wavelengths. An electron transfers electrical energy at the speed of light but does not travel at the speed of light itself; much like air or water molecules that oscillate moving particles at the speed of sound in the form of waves do not themselves move at the speed of sound.

Energy transferred from one electron to another in a wire carrying alternating current moves back and forth at the speed of light at a specific rate (the electrons themselves remain localized in the wire). The field surrounding the moving electrons generates a disturbance in the form of a wave that oscillates back and forth outside of the wire and is dispersed out into space in all directions as electromagnetic energy known as radio waves. This wave must have something waving much like air or water particles do.

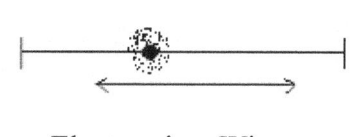

Electron in a Wire

The electrons themselves can stay relatively confined and only pass the energy from one electron to the other at the speed of light. Electrons generate electromagnetic radiation by jumping "back and forth" from one shell of an atom to another while orbiting around atoms or when jumping from one atom to another. Notice the "back and forth" motion? Remember, the "five ball pendulum", where there were five or so balls hanging on strings lined up in a row next to each other? Energy is transferred from one ball to another with very little movement of each ball; only the first and last balls really move at all. The same is true with electrons and the transfer of their energy and may also be true for energy traveling as light waves through this medium that envelopes all things.

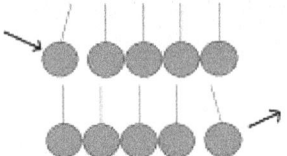

Five Ball Pendulum

Wave patterns not only are evident in the fore and aft movement in the direction of propagation, but in all directions, which bring us to the phenomenon of polarization. All particles find an equilibrium or neutral state when they expel all energy and come to rest. As energy passes through these particles however, they oscillate back and forth like pendulums, up and down, left and right or fore and aft. This back and forth waving of energy is another indicator that light moves in waves.

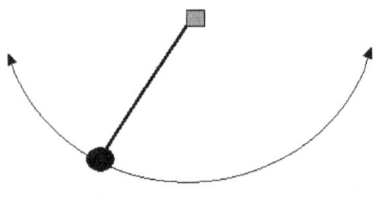

Pendulum

In the electromagnetic realm, this direction of wave motion is called polarity and is evident in all wavelengths of electromagnetic energy. The use of polarized sunglasses and the vertical or horizontal positioning of radio antennas are all examples of common utilization of the polarity phenomenon.

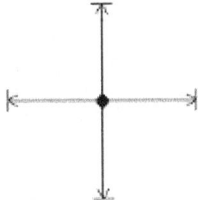

Wave Dimension

For example as light waves travel through certain crystals such as tourmaline, the crystal formations form tunnels that only certain wavelengths of light with a specific polarity will pass (Polkinghorne, pg. 18).

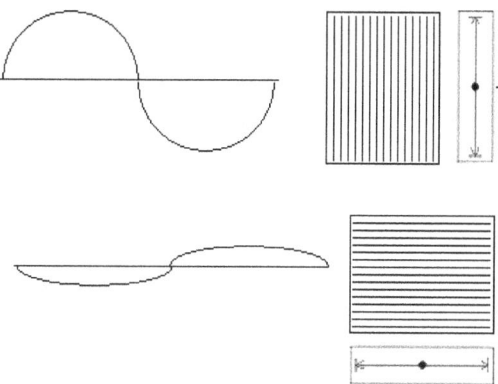

Vertical and Horizontal Polarity

Waves with opposite polarities become blocked such as vertically polarized waves trying to pass through a horizontally polarized pane of glass.

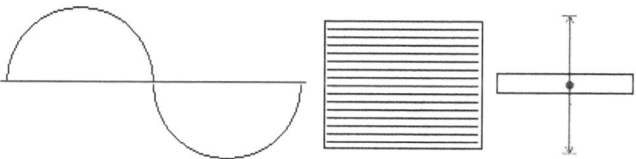

Opposite Polarity

All materials molecularly are constructed differently and have various responses to the electromagnetic energy. Some reflect, some absorb and some pass the energy straight through them. When absorbed, most of the radiation is then re-radiated back out at other wavelengths such as infrared radiation.

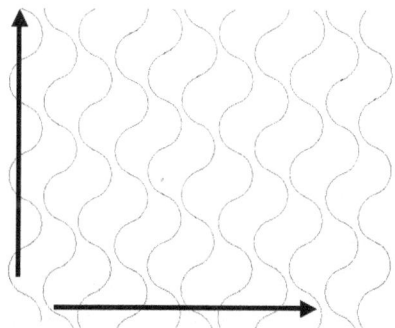

Electromagnetic Transverse Waves

Particles move back to their natural state passing energy to the next particle in waves called plane waves or linear waves. Some waves not only move energy in the direction of the wave but also perpendicular to the wave in what is known as a transverse wave – these types of waves are only possible in **solids**. Energy in the electromagnetic realm moves as transverse waves at the speed of light in all directions as seen in the previous diagram. This in turn suggests that the universe and all that we know exists as energy passing through a medium of some sort of elastic solid. The density and temperature of this medium varies creating the variations in gravity, time and space.

Waves or Particles

In the year 1900, as previously discussed, German physicist Max Planck hypothesized that energy emitted or absorbed by an atom can only be transferred in distinct quantities, which he called "quanta", Latin for "how much". This "quanta" soon played a critical role in the new study of quantum mechanics as it required that oscillating atoms only possess the energy values of E or a whole number multiples of that value. The formula for $E = h \times f$, where Planck's constant "h" is multiplied by the frequency "f", of the oscillating body (Bord, pg. 374).

The rationale behind this formula is that one full oscillation of a particle; let's say an electron, during a specific time frame, would use a specific amount of energy. Doubling the oscillations during the same time period would take twice the amount of energy and tripling the oscillations would therefore take three times the energy and so on. This oscillation is the frequency at which the electron vibrates and as it turns out is directly proportional to wavelength of the light waves emitted from an oscillating electron, whether it is from a star or a light bulb.

Planck's constant, "h" is measured in Joules (symbol J), the International System of Units (SI) energy measurement for heat, electricity and work. Planck's constant, $h = 6.63 \times 10^{-34}$ Joule seconds (Bord, pg. 375).

There are two oscillating bodies or particles that are commonly referred to as energy carriers in the quantum realm, the electron and the photon. Each of these particles can be scientifically considered as either particle or wave in nature as mathematical formulas have been developed that seem to indicate them as both.

The neutron, outside of the nucleus has a very short life span of only a few minutes, breaking down, believe it or not into a proton and an electron in a radioactive decay process known as beta decay. Even within the nucleus, this radioactive decay occurs naturally in some elements, as in carbon 14, where an electron is ejected from the nucleus at a high rate of speed and a neutron spontaneously converts into a proton; a smaller particle known as a neutrino is also emitted as a release of energy (Bord, pg. 417). Carbon 14's nucleus contains 6 protons and 8 neutrons and converts into Nitrogen 14 with 7 protons and 7 neutrons when the neutron goes through beta decay and becomes a proton. Many other short-lived elementary particles have been discovered in recent years, bringing to light the complicated constructs of the subatomic world. These include leptons and quarks, the building blocks of electrons and protons, as well as antiparticles, which mimic other atomic particles inversely as if viewed in a mirror having opposite charges and spins. The contact of a particle and anti-particle will result in the annihilation of both particles in a large burst of energy (Bord, pg. 457).

Elementary particles are the building blocks of the universe and are the fundamental constitutes of matter, either bosons or fermions depending on the type of spin (up or down). Particles normally linked with the construction of matter are called fermions and have a half-integer spin (odd integer (number) spin +1/2). Fermions are divided up into twelve categories called "flavors" and three other groups called "colors" (Bord, pg. 479). Particles associated with fundamental forces - the strong, weak and electromagnetic and possibly gravitational forces are called bosons, named for Satyendra Nath Bose, having an integer or whole number spin (Bord, pg. 458).

Fermions:

Quarks - up, down, charm, strange, top, bottom
Leptons - electron neutrino, electron, muon, muon-neutrino, tau and tau-neutrino

Bosons:

Gauge bosons - gluon, W and Z bosons and photon
Higgs boson and the theoretical graviton

Only four charged subatomic particles exist that are stable particles, the electron, proton, positron and antiproton (Asimov, pg. 246). About 500 other particles have been identified in experiments using particle accelerators in recent years, all of which are unstable and exist only for short periods of time. Subatomic particles are composite particles created out of fundamental particles such as quarks and gluons that have no known substructure. These particles can only exist under certain conditions, known as "confinement" such as in the neutron, which decays into a proton and an electron in a matter of minutes outside of an atoms nucleus (Bord, pg. 417). Fundamental particles such as leptons and quarks are highly unstable and can only exist in confinement, attached in pairs or triplets by the "strong force"; these elementary particles in turn are the main components of the electron (a lepton), proton (two up quarks and one down quark) and of course the neutron.

Although these particles may seem to exist as individual masses, they actually may simply exist as waves in the field of fundamental particles that fill the vastness of empty space. As mentioned earlier, subatomic particles possibly only exist as a quantity of these smaller fundamental particles under certain conditions. As energy moves through these particles they temporally coalesce in a form of phase transition, passing energy from one particle to another in much the same way as do fans executing the "wave" in the stands of a sports event. Since energy cannot be created or destroyed (Law of the Conservation of Energy), this energy continues on in waves from particle to particle. This energy at the quantum level in the form of Plank's Constant also cannot be divided, transferring all of its energy to the next particles.

Electron as Particles Temporally Coalesced

Under specific conditions, this pressure wave possibly becomes trapped in a continuous circular motion, confined to a specific pressure gradient around an atom where it alone can exist, manifesting itself as an electron. As energy is absorbed into this wave, the wave changes momentum and instantaneously repositions itself in a pressure gradient where different conditions allow the "particle" to reappear as an electron in another shell or pressure gradient; such as transitioning from orbit 5 to 8 at a new velocity. Think of this as fans performing the wave around the stadium only in row 55, then halfway around stopping and shifting instantly to only row 75 for the rest of the way around. Although there are people in every seat, the wave manifests itself as people stand and raise their hands one after another and then sit down in a continuous fashion. The distance the wave has to travel as it circles the stadium varies from row to row as well time it takes to complete one circle.

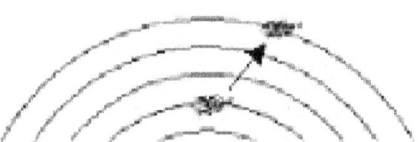

Transition from Orbit 5 to 8

If we look at the electron as a single particle in orbit, as if a satellite around the atomic nuclei, the electrons wavelength of the smallest orbit of the hydrogen atom with a velocity of the $2.19*10^6$ m/s (meters per second) is: (λ is wavelength)

λ = h/mv (Planck's constant and Momentum), then:

λ = 6.63 X 10^{-34} J-s/$1.995*10^{-24}$ kilogram-meters per sec.

or:

λ = 3.32 X 10^{-10} meters or .332 nm (nanometers)

Nanometer = one billionth of a meter (0.000000001)

If r is the radius of an orbit then the circumference of that orbit is = $2\pi r$, which is also its wavelength.

Since $2\pi r = \lambda = .332$ nm then the radius of the smallest orbit or orbit 1 is:

1 radius = $\dfrac{.332 \text{ nm}}{2\pi}$ = $\dfrac{.332 \text{ nm}}{2 \text{ X } 3.14}$ = $\dfrac{.332 \text{ nm}}{6.28}$ or .0529 nm

Orbit 2 would theoretically then be 2λ or

= 2 X .332 nm = .664 nm.

The circumference of the next orbit or orbit 2 is:

2 radius = $\dfrac{.664 \text{ nm}}{2\pi}$ = $\dfrac{.664 \text{ nm}}{2 \text{ X } 3.14}$ = $\dfrac{.664 \text{ nm}}{6.28}$ or .106 nm

Orbit 3 would then be 3λ or = 3 X .332 nm = .996 nm.

The circumference of the next orbit or orbit 2 is:

3 radius = $\dfrac{.996 \text{ nm}}{2\pi}$ = $\dfrac{.664 \text{ nm}}{2 \text{ X } 3.14}$ = $\dfrac{.996 \text{ nm}}{6.28}$ or .159 nm

Orbit 1 radius = 1λ = .0529 nm

Orbit 2 radius = 2λ = .1057 nm which is 0.0529 nm further

Orbit 3 radius = 3λ = .1586 nm which is 0.0529 nm further

Orbit 4 radius = 4λ = .2115 nm which is 0.0529 nm further

Orbit 5 radius = 5λ = .2643 nm which is 0.0529 nm further

Energy = h (Planck's Constant) X Frequency

Frequency = $\dfrac{\text{Speed of light}}{\text{Wavelength}}$

Wavelength = $\dfrac{\text{Speed of light}}{\text{Frequency}}$

Speed of light = c   Frequency = f   Wavelength = λ

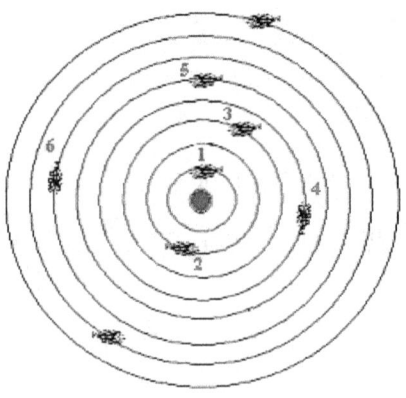

Waves of Particles

We see that if we increase each orbit by multiples of the wavelengths, each orbital radius is increased proportionately as in the previous illustration, 0.0529 nm further out for each new orbit. This is not the case however; each orbit changes according to variations in energy in fixed quantities of multiples of *h*, Planck's constant, not fixed distances from the nucleus. These orbits are not measured in radius or circumference, but in Electron Volts. Each orbits radius decreases slightly the further out in orbit the electrons are. As an electron transitions from one orbital or shell to another, the distance between orbits also changes.

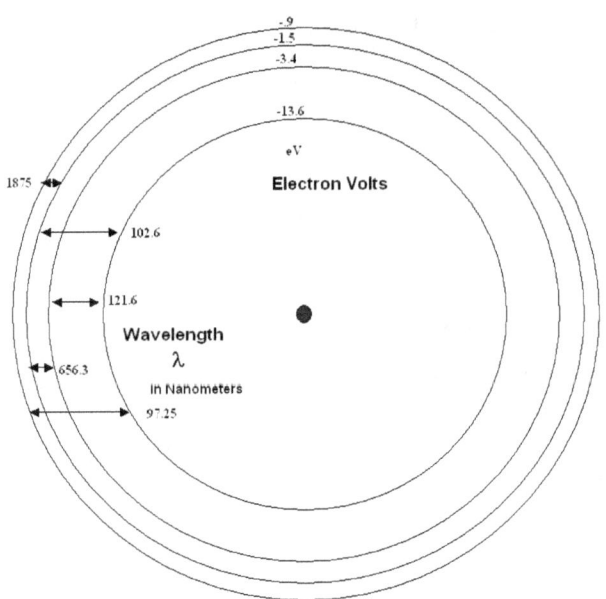

Radius Change in Orbit

The difference between orbit 1 and orbit 3 is greater than the difference between orbit 4 and orbit 6. The velocity of the electron also varies from orbit to orbit. As the distance of the electron from the nucleus increases, the force that pulls the electron towards the nucleus decreases by the square of the distance. This phenomenon is common to many processes such as the change in gravity, magnetism and the change in intensity of electromagnetic energy or light as the distance from the source increases. This in turn has a direct effect on the electrons velocity.

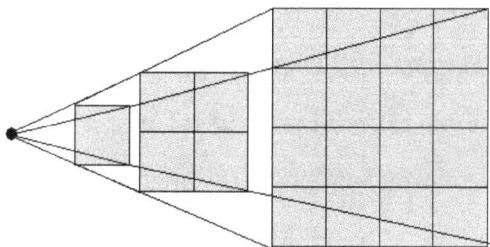

Attraction Decreases by the Square of the Distance

If an electron has the mass m, and the nucleus has mass M, and the electrons distance from the center of the nucleus is r, then g is the force that the nucleus exerts on the electron or g = $Gm M /r^2 /m$ (where G is a universal constant for the attraction of mass). If the electron would fall towards the nucleus, the nucleus's attraction will cause it to accelerate toward the center of the nucleus. According to Newton's second law (F = ma), this acceleration g must equal:

$$g = (Gm M /r^2)/m, \text{ or } g = GM/ r^2$$

For the electron as a single particle, the proportionate change in energy input does not produce a proportional change in orbit. The velocity of the electron varies according to the change in the nucleus's attraction due to the change of distance in radius from the nucleus.

The angular momentum (p) of each orbit or mrv is:

$$F = \frac{mv^2}{r} \qquad \frac{v^2}{r} = \frac{GM}{r^2}, \text{ or } v = \sqrt{\frac{GM}{r}}$$

(Mass x Radius x Velocity)

$$mrv = \frac{h}{2\pi}$$

Therefore the angular momentum for:

Orbit 1 = mrv = $\dfrac{6.63 \times 10^{-34}\,\text{J-s}}{6.28}$

Planck's constant can be looked at in two ways:

In Joule/ seconds = 6.63×10^{-34} J-s or

Electron Volts (eV) = 40136×10^{-15} eV/ Hz

Angular Momentum (mrv) therefore = 1.0552×10^{-34}

Velocity in the lowest orbit is 2,189,638 m/s

Mass of an electron of 9.11×10^{-31} Kg

Velocity in the lowest orbit is 2.19×10^{6} m/s.

Radius at lowest orbit = 5.29×10^{-11} meters

Wavelength = λ = .332 nanometers

Momentum (p) would be: 1.995×10^{-24}

The mrv = 1.0552×10^{-34}

If we go up to $2h$ or the next energy level up we get:

mrv = $\dfrac{2h}{2\pi}$

If the radius were twice the ground state:

Radius = r = $2 \times 5.29 \times 10^{-11}$ or 1.06×10^{-10}

Velocity would be = 2,189,638 m/s

Mass of an electron = 9.11×10^{-31} Kg

Momentum would be = 1.995×10^{-24}

Angular Momentum = 2.1104×10^{-34}

If we stay at 2*h*:

$$mrv = \frac{2h}{2\pi}$$

And the radius was three times the ground state:

Radius = r = 3 X 5.29 X 10^{-11} or 1.59 X 10^{-10} (increased)

Velocity would be = 1,459,758 m/s (decreased)

Mass of an electron = 9. 11*10^{-31} Kg

Momentum would be = 1. 330 X 10^{-24} (decreased)

Angular Momentum = 2. 1104 X 10^{-34} (same)

Looking at the electron as a particle, we see that at the same level of energy, 2h, we can have different velocities and different linear momentums. As far as the electron goes, the higher its orbit is the lower its velocity and the lower the linear momentum. Likewise, the lower the orbit of the electron, the higher its velocity is and the higher its linear momentum is. Angular Momentum (mrv) must therefore be calculated in order to show the total energy of the electron.

We also see that attraction forces change according to variations in orbits; the higher the orbit, the lower the attraction forces are. This change in distance has a direct effect on the velocity and linear momentum of an electron. Also, the higher the energy of the electron, the further out in orbit an electron will go; but the further out the electron is in orbit, the slower it travels and the less linear momentum it has. What then creates the fixed wavelengths of energy emitted from these systems? How do these slower electrons create the higher energy photons of shorter wavelengths? It seems the longer wavelength of the electron, the shorter the wavelength of the photon it creates – a paradox of sorts.

When sound is created from a guitar string, the string oscillates back and forth at a specific rate every second. Its wavelength is equal to the length the string physically moves, first one direction and then the other. The longer the wavelength of the strings oscillations, the lower the frequency is and the less energy is being used; the shorter the wavelength, the higher the frequency and the more energy used. To double the frequency of the strings oscillation is to double the energy radiated by the string at the same amplitude. This strings motion disturbs the air around it and the energy it radiates is dispersed into the air as compression waves or plane waves. Each of these plane waves travels away from the string at the speed of sound in that medium (air) at fixed wavelengths.

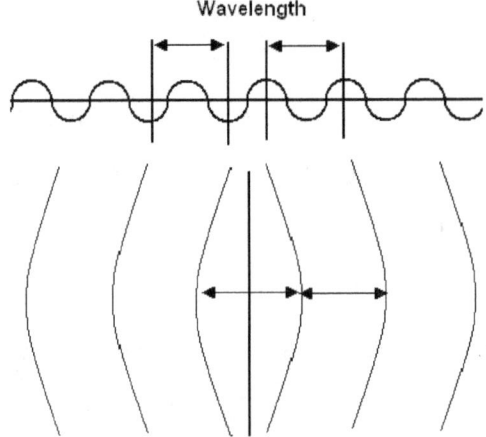

Guitar String Oscillations

An electron likewise oscillates back and forth as a guitar string. It does this in two ways, during its normal orbit around the nucleus of an atom in a circular motion at multiple levels and as it momentarily jumps from one orbit to another. The circular motion of the electron in orbit around the nucleus traveling a full 360 degrees is one full wavelength with each change in orbital radius a different wavelength. The velocity of the electron in a specific orbit produces a repetitive motion that repeats thousands of times per second, measured in Hertz (Hz). The faster the electron oscillates, and the smaller its orbit, the higher the frequency of the electron. The larger the orbit of the electron is the slower its velocity and the lower the frequency of the electron. The wavelength of the electron is dependent on its velocity and orbital circumference, not just the diameter of its orbit. One thing to keep in mind is that although an electron in a higher orbit is slower than lower orbiting electrons, it has **more energy** due to angular momentum. Faster electrons in lower orbit likewise have **less energy**.

Velocity in the lowest orbit is $2.19*10^6$ m/s

Radius at lowest orbit = 5.29×10^{-11} meters

Circumference = $2\pi r$ = or .332 nm

$$f = \frac{v}{\lambda} \qquad v = f\lambda$$

Orbit 1 = f = 6.59×10^{15}

Orbit 2 = f = 2.19×10^{15}

Orbit 3 = f = 1.098×10^{15}

Mass	9.11×10^{-31}	9.11×10^{-31}	9.11×10^{-31}
Radius	5.29×10^{-11}	5.29×10^{-11}	5.29×10^{-11}
Velocity	2,190,000	1,459,712	1,094,783
Wavelength	3.32212×10^{-10}	6.64424×10^{-10}	9.96636×10^{-10}
Hz	6.59218×10^{15}	2.19696×10^{15}	1.09848×10^{15}
mrv	1.0554×10^{-34}	1.40693×10^{-34}	1.58279×10^{-34}

Orbital Relationships

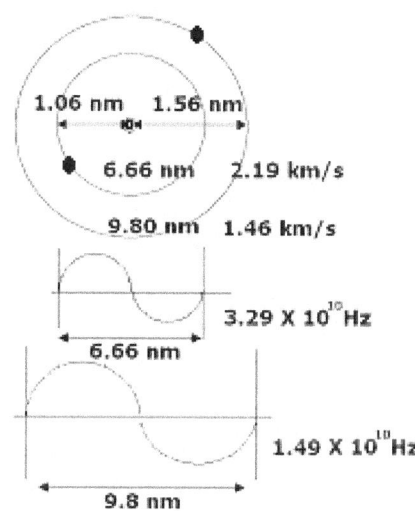

Wavelengths of Two Orbits

An electron also is constantly changing orbit as it absorbs energy bombarding it from all directions by photons. Each time the electron absorbs energy from a photon, it flies further out in orbit, and then releases the energy it absorbed and retreats back to a stable orbit. Each of these jumps or quantum leaps of the electron absorbs or emits a photon in fixed energy bundles previously discussed.

What is different here is that unlike the guitar strings wavelength, which is proportionate with the strings vibrations, the energy released from an electron works in an inverse way. The larger the movement of the electron, whether in orbit or during a quantum jump, the shorter the wavelength of energy is that is emitted. Likewise, the shorter the jump or smaller the orbit, the longer the wavelength of energy released. The answer to this dilemma may be in that the time for an electron to jump from here to there, no matter how many shells it may jump, is always the same; shell three to four or one to seven. If a constant say z for time is divided by the energy h that each shell releases (one packet for each shell jumped) then a three to four jump would be z/1h the for the longest wave length of x and a one to seven jump would be z/6h for a shorter wave length of y.

As covered earlier, a particle can possibly be viewed as a shockwave whose velocity is limited by the density and temperature of the material or medium the electron wave is flowing through. When viewing an electron in this matter, the closer to the nucleus an electron wave moves, the denser and cooler the medium is that the electron wave is moving through (since the particle itself may be a condensate of these cooled particles in wave form). .

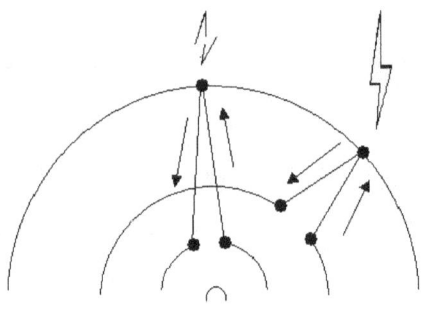

Quantum Leap

The further out that the electron wave moves from the nucleus, the hotter and less dense the medium theoretically becomes in which the wave is moving through. The electron as a fixed wave bundle of energy can only exist as a shock wave in one region at a time, at a specific velocity. If energy were added as a second fixed bundle of energy to the first wave (another photon wave is merged with the electron wave), the momentum would then be doubled (2h). With the momentum doubled but the mass of the combined particles in the wave still the same, the angular momentum and the orbit of the electron therefore must change. However, the present position of the wave in the medium it is presently moving through can no longer exist at this higher energy level, forcing the wave pattern to reform as a shockwave further away from the nucleus in a less dense region. This new wave then travels in a circular pattern around the atom further out staying within the new density bounds at a different velocity. This wave energy will be created at a higher angular velocity (mvr), which is actually a lower linear velocity but at a higher energy level. We must remember here that the electron itself is not a particle moving the distance of low orbit to high orbit or vice versa, it is simply energy dissipating in one level and instantaneously reforming in another level in steps. This produces the various wavelengths of light emissions that vary according to temperature and the element emitting the electromagnetic radiation.

We can again picture this as fans doing the wave around a stadium only in row 55. The more people in the row generating the wave, the longer it takes for the wave to move around the stadium. Inner rows have less distance to travel around the stadium but may have more people packed in the seats, slowing down the wave. Seats near the back of the stadium have many more seats for the wave to move through but possible has less people in the seats changing the velocity of the wave as well.

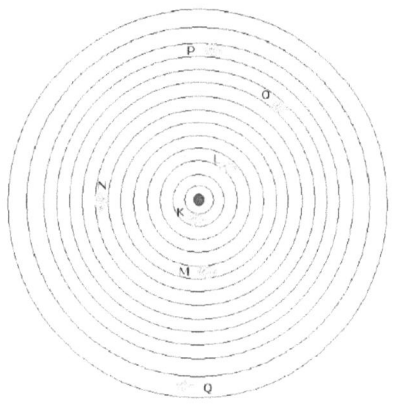

Temp in Kelvin	Velocity of Light
K 10.13	2,189,856
L 20.46	4,421,846
M 41.32	8,928,773
N 83.44	18,029,345
O 168.49	36,405,593
P 340.21	73,511,667
Q 686.97	148,437,772
∞1387.16	299,731,634

Velocity vs. Temperature

The previous illustration shows the seven shells that an electron wave can move through. The illustration shows how a possible temperature and density change of a medium from one shell to another affects the velocity of a shock wave through that medium. The velocity of an electron above Q, the highest orbit, breaks the bonds of the nucleus at the speed of light c, 300,000 km/sec, an escape velocity if you will, sending energy out through open space in the form of a photon in a straight line.

Sound

Generally speaking, sound travels slower in denser materials. The speed of sound within one material such as nitrogen or water however, is solely reliant on the temperature of that material, not the materials pressure or density. Sound travels faster however in solids and liquids than gases, because the molecular bonds of solid materials are much stronger than that of gases. Solids will have a higher sound speeds than liquids, and liquids will have a higher speed than gases (Bord, pg. 209). The speed of sound in air at standard temperature and pressure (25°C or 59 ° F) is 343 meters per second (767 miles per hour). At 0° C or 32 ° F the speed of sound in air is 330m/s (740 mph), 13 m/s slower. The speed of sound in water is about 1,500 meters per second (3,355 miles per hour) (Bord, pg. 224).

The wavelength of the sound wave at 240Hz is

$$\lambda f, \ \lambda = v/f = (340m/s)/ (264 /s) = 1.29 \text{ meters}$$

The speed of sound depends on the temperature, the density of the material (mass per volume) and the elasticity of the material or *medium* it is traveling through. As sound transitions from air to a solid, the speed of sound increases as well as its wavelength. As sound transitions from a solid to air, the speed of sound decreases as well as its wavelength.

Sound through Air - Wavelength 1.3792 Meters

Sound through Cork - Wavelength 2.0833 Meters

Sound through Alcohol - Wavelength 5.1667 Meters

(Sea Level)	M/Sec	Formula	MPH	Freq. Hz	W/L in Meters
Air (0° C. /32° F.)	331	0.44704	740.426	240	1.3792
Cork	500	0.44704	1,118.468	240	2.0833
Alcohol	1,240	0.44704	2,773.801	240	5.1667
Water	1,500	0.44704	3,355.404	240	6.2500
Brick	3,650	0.44704	8,164.817	240	15.2083
Wood (Oak)	3,850	0.44704	8,612.205	240	16.0417
Glass	4,540	0.44704	10,155.691	240	18.9167
Aluminum	5,000	0.44704	11,184.681	240	20.8333
Iron	5,103	0.44704	11,415.086	240	21.2625
Steel	5,200	0.44704	11,632.069	240	21.6667

Sound Variations in Media

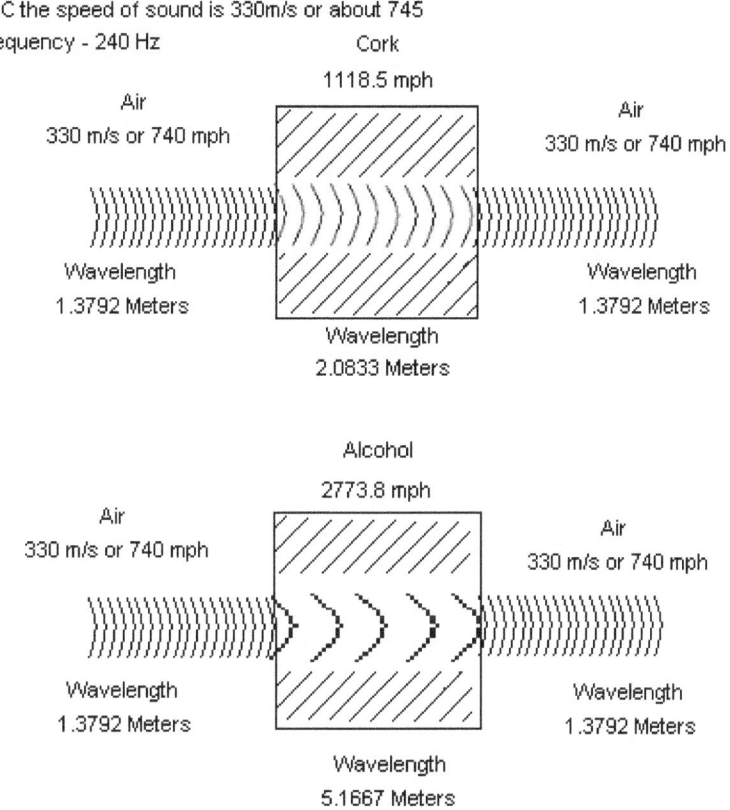

Light

As light transitions from a vacuum to a new *medium*, or from less dense *medium* to a greater density *medium*, the speed of light decreases and its wavelength also decreases.

As light transitions from a *medium* to a vacuum or to a less dense *medium*, the speed of light increases as well as its wavelength.

Light through <u>Air</u> - Speed 185,226.52 Miles/ Sec.

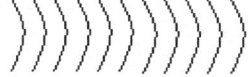

Light through <u>Water</u> - Speed 139,746.73 Miles/ Sec.

Light through <u>Crown Glass</u> - Speed 114,989.13 Miles/ Sec.

Medium	Speed of Light	Frequency in Hz	Wavelength in Nanometers	Percent of Light
Vacuum	186,282.40	4×10^{14}	680.00	100.00%
Air	185,226.52	4×10^{14}	676.15	99.43%
Water	139,746.73	4×10^{14}	510.13	75.45%
Glass	114,989.13	4×10^{14}	419.75	62.08%

Light Variations in Media

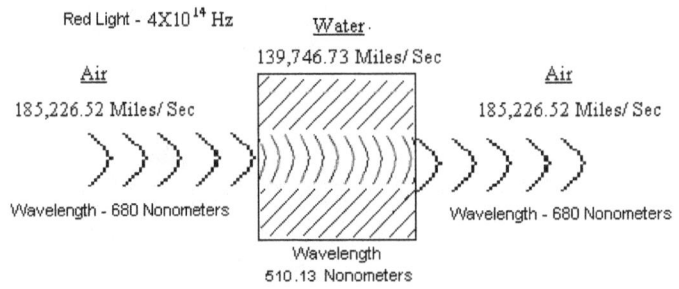

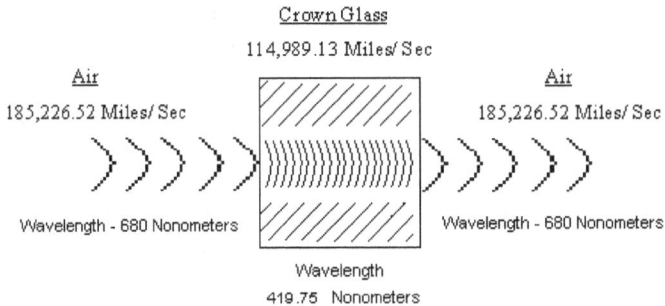

Electromagnetic Waves

In the vacuum of space, absent of any interference of fields, the velocity of electromagnetic waves as energy is said to be a constant 300,000 km/sec or 186,000 miles/sec. In denser materials such as water and glass, the propagation is slower.

The simplest way to envision electromagnetic energy is to see how radio waves are propagated. If the length of an unshielded wire is proportionate to the wavelength of the cycle of electrons in an alternating current, electrons in a wire influenced by this electric current will generate an electromagnetic wave radiating out in all directions. Heinrich Hertz was the first discovered radio waves in 1888 (Asimov, pg. 51)

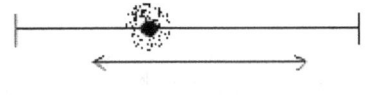

Alternating Current

An alternating current is a flow of electrons in a conductor whose current alternates back and forth, positive then to negative, at a specific frequency or rate per second. The frequency of the electromagnetic energy or field generated by this current directly corresponds with the frequency of the alternating current. This shifting of the electromagnetic field can be sensed for miles depending on the strength of propagation and is dispersed as an oscillating electromagnetic wave in all directions. This energy possibly begins with the field of particles surrounding the electrons moving in the wire, which in turn disturb the particles in empty space around the wire as the electrons oscillates back and forth. Electromagnetic energy would then be transferred from one particle to another in this field of particles – which we will call the *medium*. This field of particles or *medium* is what will do the waving and possibly is the substance of which all things are made. The velocity of this transfer of energy depends on the density of the *medium*, 300,000 km/sec (186,000 m/sec) in a vacuum for example. All electromagnetic energy would then be the back and forth oscillation of this *medium* as waves at different frequencies.

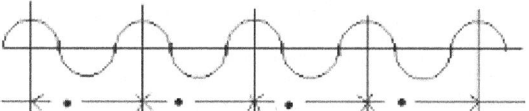

Wave Patterns

The transfer of this energy is at the speed of light. The wavelength is determined by the frequency of the oscillations, the higher the frequency the shorter the wavelength. The sine wave shown above depicts the electric current flow with the solid line representing no current flow, rising to a peak voltage or amplitude and decreasing back to zero volts. The sine wave then progresses below the line, representing the electrons in the wire flowing in the opposite direction to peak amplitude and then back to zero. These cycles continue thousands or even millions of times per second; electrons first flowing one way, decreasing in amplitude/ energy and then back in the other direction to full amplitude. The wavelength of electromagnetic radiation is measured as the distance between to peaks or troughs (valleys) in the cycle. These frequencies range from very low radio frequencies, to visible light, to x-rays and gamma radiation.

Some waves such as in radio waves are transmitted in half waves in one direction as in pulse and single sideband radio. This may be similar in all wavelengths of the electromagnetic spectrum in various transmission forms, both naturally and manmade.

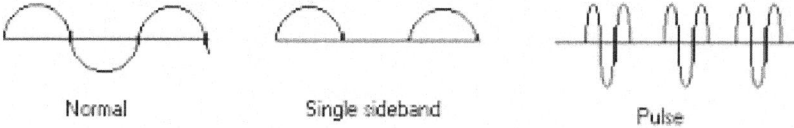

Normal   Single sideband   Pulse

Radio Transmission Patterns

With the fore and aft motion of the *medium,* we see that combining waves in phase, or combining waves out of phase, may interfere with their propagation.

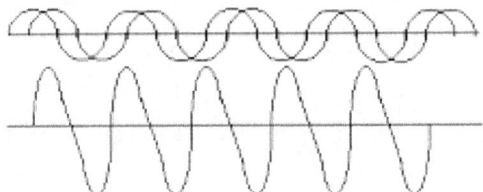

Combined Waves

As seen above, two peaks combined in phase amplify the peaks and the troughs.

Two peaks combined 180 degrees out of phase; cancel each other out leaving a flat line.

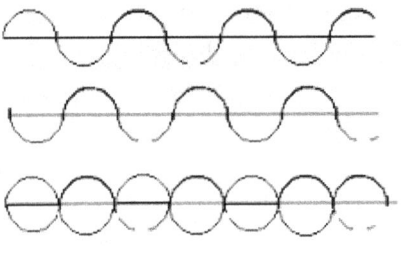

Cancelled Waves

This phenomenon exists in all ranges of the electromagnetic spectrum as well as sound waves in air and also with waves in water. Different materials react to different wavelengths of the spectrum, such as wave-guides (microwave antennas), radio antennas, photographic paper, x-ray sensitive film, and so on. From these materials we have created many different ways to detect and produce various wavelengths of electromagnetic energy for all sorts of things such as manufacturing, communication and recreation.

Doppler Effect in the electromagnetic spectrum is a shift in the frequency produced or reflected by moving objects, similar to the very noticeable change in sound waves as earlier noted with Einstein's train. With sound, as an object such as a train moves towards an observer, the sound the object produces seems to change in pitch as it passes. The frequency or pitch emitted from the object however is constant but is perceived to be higher as it approaches the observer. The movement of the object compresses the waves transmitted in front of it, causing a perceived change in pitch. Likewise, as the object passes the observer, the wave emitted from the object becomes stretched causing the observer to perceive the sound to be lower in pitch (Ditchburn, pg. 37). The same occurrence takes place with light waves and radio waves.

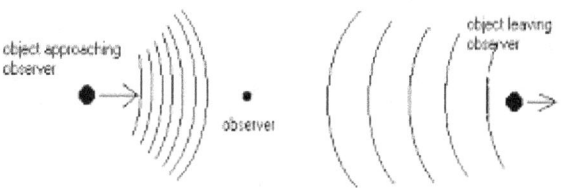

Doppler Effect

Technologies such as weather radar use this phenomenon to detect wind shear and possible tornado activity. Another use is the radar gun, utilized by police officers to detect speeders on the highway. Gravitational forces or movement of objects that emit or reflect light as covered earlier produce a Red Shift or a Blue Shift in light waves.

The Electromagnetic Spectrum

Imagine the *medium* as a fine gas like particle that fills all things, surrounds all things and possibly is what all things are composed of. In the electromagnetic spectrum, an emitter, possibly a star, radio antenna or light bulb, stimulates or disturbs the *medium* and energy is passed from one particle of the *medium* to another. The velocity of the transfer of energy through the *medium* is relatively constant, and dependent only on the density and temperature of the *medium*. No matter what the velocity of the emitter toward the receiver or the velocity of the receiver towards the emitter, the velocity of the energy through the stationary space (*medium)* between the objects is always the same.

Even though the transfer of the energy through the *medium* or space is constant, the Doppler effect is still noticeable. If either the receiver or emitter or both move away from each other, the wavelength is stretched and perceived to be longer and a change in the light analyzed shifts towards red light, known as red shift.

Red Shift

If either the receiver or emitter or both move towards each other, the wavelength will be compressed and perceived to be shorter. Within the light spectrum, a change in color is noticed, in what is known as blue shift. In any case, the wavelength emitted is actually the same. It is either that the wave as a whole is compressed or stretched by the moving emitter or receiver that causes the perception of waves to be compressed or stretched.

Blue Shift

As mentioned before, the velocity of the transfer of energy seems to be the same from one particle of the *medium* to another. The denser or closer together the particles of the *medium* are together, the longer it seems to take for energy to be transferred.

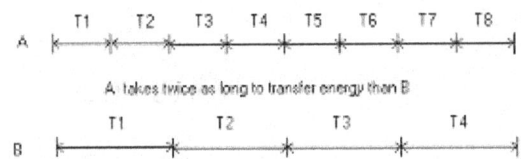

Velocity Through the *Medium*

As reviewed previously, energy is expressed in the quantum world in "packets" or "quanta" which means in Latin, "how much." This is also known as "Planck's Constant" with the formula $E = hf$ where E equals total energy, "h" is Planck's constant and "f" is the oscillation or frequency of the electromagnetic energy (Polkinghorne, pg. 6). In the figure above, we show that it takes "T" amount of time to move one particle of the medium by one "packet" of energy into the field of the next particle, causing a chain reaction. All energy is passed from one particle to the next, developing a continuous chain reaction; this "packet" of energy cannot be split up. According to the "law of conservation of energy", no energy can be gained or lost. This is similar with electrons, as we will soon see.

Let's presume that each particle of the *medium* has a specific mass and inertia. It takes a specific amount of energy to move that particle from at rest to motion. Likewise, the *medium* particles take the same amount of energy to be returned back to rest. The higher the frequency, the more energy needed to start and stop each particle in a given period of time.

The question now arises as to how the velocity of light always remains the same to the observer. Light is always measured to be 300,000-km/ sec or 186,000 m/sec in a vacuum. How can this be? Einstein's whole basis for the theory of relativity is based on this discovery. He hypothesized that an increased velocity shifts space and time. As velocity increases, a clock starts to change time and mass begins to accumulate.

This hypothesis generates many scenarios, most of which possibly are only perceptions of the observer and not reality, similar to what we saw with the train and the embankment. For example a clock on a rocket launched from Earth may seem to be ticking slower to an observer. This may only be perceived because of the rate of speed, i.e. the Doppler Effect. Likewise, as the rocket returns, the clock may be perceived as ticking faster because of the Doppler Effect, but in reality, no time was lost or gained on the trip. A sphere traveling close to the speed of light going from left to right of a stationary observer may be perceived by the observer to be compressed (Born, pg.319). Light would be absorbed and reflected off of the leading edge of the sphere. The sphere actually will run into the light reflected off the leading edge and make it appear to be compressed. Does this affect its reality? No, only in the way it is perceived, just as the sea bottom is distorted through the waves to the viewer on a boat.

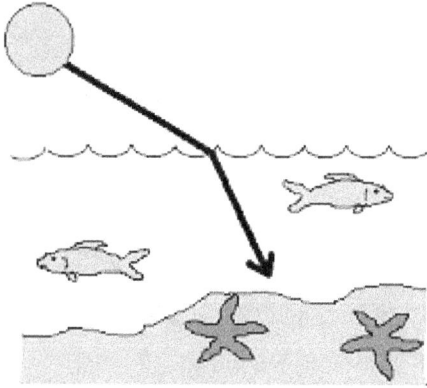

Distortion

Now let's look at the perceived speed of light to an observer. The *medium* in empty space is stationary and each particle is equidistant from each other, that is, in the natural "pressure" of the *medium*. Moving objects however and gravitational forces can compress or expand the *medium* causing higher and lower pressures or densities from its natural state. As an object moves toward a source of light at a constant velocity, the *medium* particles in line with both objects will compress. The number of particles will remain approximately the same between the emitter and the object, and only slowly change.

Imagine a tennis ball shot through a tube with approximately the same inside diameter as the diameter of the ball. Both ends of the tube are closed. The number of particles of air between the end of the tube B towards which the ball is moving, and the ball, remains the same except for a few that squeeze by the ball, with the air in the tube being compressed with increased pressure. Likewise between the aft end of the tube and the ball the number of particles remains the same even though it is being expanded and the pressure is decreasing.

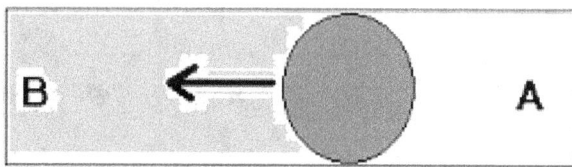

Compression of the Medium

Now imagine traveling towards Jupiter in a spaceship near the speed of light. Light emitted from Jupiter cannot travel faster than the speed of light relative to the observer in the spaceship. Just as with the ball in the tube, light waves would have of be compressed in front of the spaceship bending space-time in front of the ship and stretching out behind the ship. Theoretically, the spaceship could be much closer to Jupiter than would be perceived onboard. The question is, what changes dimensions with speed? The spaceship or space itself?

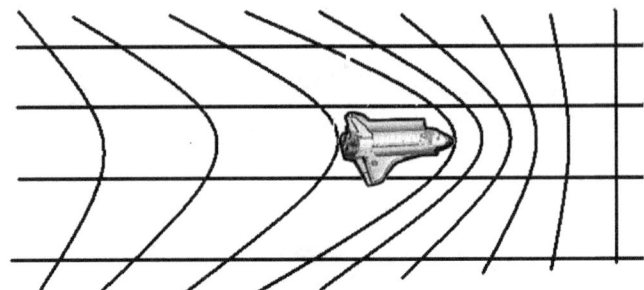

Compression of the Time and Space

Of course there is to be some difference between subatomic particles and sound through the air and water; this is because of the "law of conservation of energy". Energy transferred through air and water disperses energy into various factors, heat, motion, sound etc. In the subatomic electromagnetic spectrums, changes in energy only are seen in frequency changes of the emissions. The more energy released the higher the frequency of the radiating waves.

Now imagine all the particles of *medium* between the two moving objects. As the objects come closer, very few particles escape the pressure between the objects as it compresses. The number of particles increases in a given space as velocity increases. Therefore the speed of light changes proportionately to the change in velocity of the target to the emitter. This is relative to an outside observer. The outside observer cannot see this energy because he is outside the "compression zone". To the observer on the moving object, the speed of light has not changed no matter what his velocity. If the moving object was to stop, the pressure between the objects neutralizes and the speed of light still is measured at 300,000 km/sec. Now you say; if the speed of light is always the same to the moving observer, then there should be no apparent Doppler effect noticeable to the observer. Waves must then travel out in all directions from the source and not be affected by the compression. If this were true, the peaks of the electromagnetic waves moving towards an object would be perceived to be moving faster than light even though the light measured on the moving object is measured at the speed of light, 300,000 km/sec.

We see that the wavelength in miles is:

$\dfrac{186,000\ldots}{\text{Frequency}}$ or in ft. … $\dfrac{983,559}{\text{Frequency in kHz}}$

The wavelength is directly proportionate to the speed of emitter. How can you then have a Doppler effect with the perceived frequency of light shifting when the speed of light measured in the same wave doesn't change? This is a paradox.

One way of looking at this is to use a submarine traveling through water as an example. The sub sends a sonar beacon towards another sub in the water. The sub is traveling through the water at 30 knots. As soon as the sonar beacon exits the sub's transmitter, it moves in waves at a constant speed through the water relative to the movement of the water particles and independent of the speed of the sub. The speed of sound through water depends on the temperature of the water and is constant, in this case 3,355 miles per hour no matter what the speed of the sub. On the other hand, the faster the sub travels through the water, the shorter the wavelength and the higher the sonar frequency at the target due to the Doppler Effect. The frequency of the sonar increases proportionately with the increase in velocity of the sub. The speed of sound through the water from the sub however remains the same. As you can see in this figure, the Doppler Effect is easily noticeable.

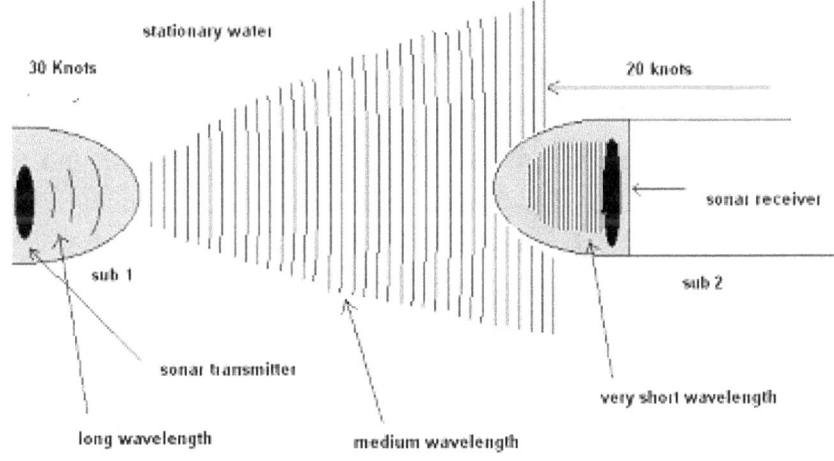

Sonar Waves

Fields and Forces

There are four fundamental interactive forces in nature. These four categories of forces studied in quantum mechanics are gravitational force, electromagnetic force, the weak force and the strong nuclear force; the weakest of these is gravity.

Gravitational force is the affect that every particle has on other particles and its force is directly proportional to the combined mass of objects within the field; this field's strength decreases by the square of the distance between the particles. For example, the sun and the earth: here the total gravitational attraction the sun has on the earth must be added to the gravitational pull the earth has on the sun. It is not just the powerful gravitational pull the sun alone that pulls on the earth but also the earth's own pull on the sun. Likewise, a man standing on the earth not only is attracted by the earth's pull on him but also his mass that is also attracting the earth. The greater the total mass (earth + sun), the greater the gravitational force that is pulling the objects together.

Long ago, Aristotle had reasoned, supposedly erroneously, that the velocity of falling objects were dependent on their weights. Galileo hundreds of years later discover that the velocity of two objects of different weights fell at the same rate; this view is still held today. In fact, Aristotle was somewhat correct. A marbles rate of acceleration towards the earth is not only caused by the earth's attraction of the marble but also the marbles attraction of the earth. The acceleration rate towards the earth is due to the combined mass of both objects, therefore, a bowling ball will have an ever so slightly higher rate of acceleration. This is due to the fact that the combined mass of the earth and the mass of the bowling ball is slightly greater than that of the combined mass on the earth and the marble.

Gravity today is believed to be manifested by a mass-less particle called a graviton. This is done through an exchange of gravitons in gravitational waves between the particles of two bodies. This force, unlike the other three forces, only attracts and does not repel. Although it is the weakest of the four forces, its effects can be generated over a very large distance (Hawking, 1988, pg. 72).

The electromagnetic force is a force that is both attractive and repelling in nature and only acts between two or more electrically charged particles. Here, like charges repel and opposites attract, positive repels positive but attracted negative. Examples are where a proton attracts an electron but repel another proton; this is the attracting force that causes an electron to orbit the nucleus of an atom that contains the protons (Hawking, 1988, pg. 73).

As protons and electrons particles interact with each other, orbiting electrons will in certain situations be shared with the nucleus of other atoms; in a sense, tying the two atoms together in what is called a covalent bond.

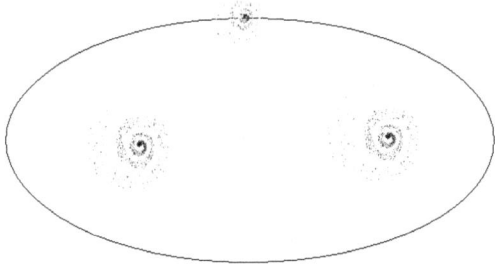

Bonding of Particles

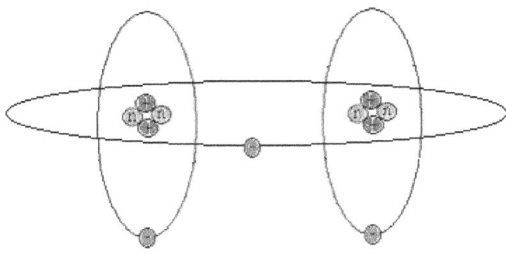

Bonding of Particles

In electrically balanced conditions, electrons and protons equal each other in numbers. In cases however where imbalances occur, electrons will try to passed from an area were excessive electrons are grouped to a region of electron deficiency. If a conductive path is available, a flow of electrons from one area to another in an electric current will occur– the principle of the battery. As these electrons move from one atom to another, a field is produced surrounding the electrons, moving with the electrical current disturbing space around them as they flow with what is commonly called an electric field.

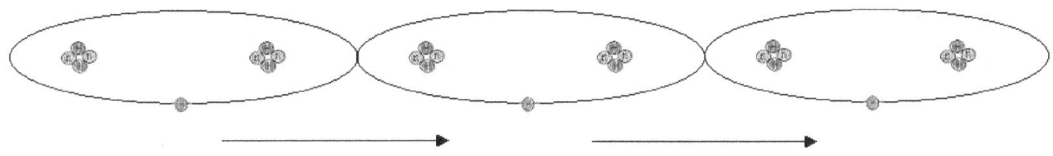

Electron Flow

Magnetic fields are caused by the motion of electrons in a circuit as in the case of an electromagnet. Here the electrical field creates an attractive or repulsive force depending on the direction of the flow; its intensity is proportional to the amount of current in the circuit. A natural magnet such as magnetite also has a magnetic field but is caused by the alignment of atoms in the substance and possible flow of the electrons naturally within the substance. The synchronous movement of electrons moving in unison and perfect alignment of the electron orbits within the magnetic material produces these magnetic fields.

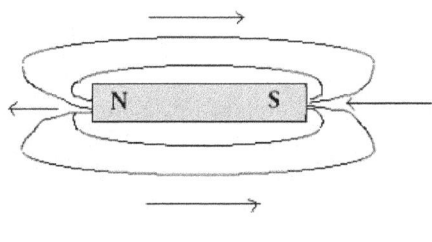

Magnet

The electromagnetic field in the form of a radio wave is also a variant of this electromagnetic force. Light waves are also electromagnetic waves that are at higher frequencies than radio waves and are manifested by the ejection of photons by the movements of electrons.

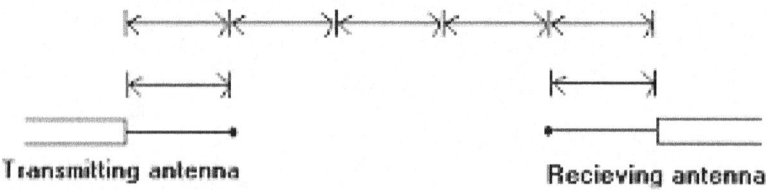

Transmitting antenna   Recieving antenna

Radio Waves

The weak nuclear force is found in nuclear radioactive decay such as beta decay, the breakdown of the neutron into an electron and a proton. Here, outside of the confines of the atom, a neutron no longer remains a neutron and breaks down due to the exchange of heavy W and Z Bosons. Particles react differently at different levels of energy, at specific energy levels symmetry can be broken causing particles to change, spontaneously breaking down (Hawking, 1988, pg. 74).

The strong nuclear force is the force that holds the building blocks of particles together. The quarks that make up protons and neutrons and the nucleus of the atom are possibly held together by the gluon. This gluon binds them together in a process called confinement, combining only specific combinations of different types of quarks (green, blue or red). Combining anti quarks and quarks may create the unstable meson of which then breaks down into an electron. The high energies of particle accelerators weaken the strong force, allowing these particles to be broken down into what appears to be "free" quarks and gluons (Hawking, 1988, pg. 76).

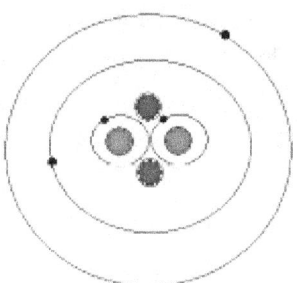

These forces are all forms of energy. Energy as we have learned is the movement or the potential movement of mass. Scientifically speaking, energy is a variable quantity that is a characteristic of an object. Energy is the ability to do work or that which is transferred when work is done (Bord, pg. G-1). There are several ways to describe energy, kinetic, potential, gravitational, electromagnetic, chemical and nuclear. Energy can be transformed from one form of energy to another such as light into heat, but cannot be created nor destroyed. This is known as the Law of the Conservation of Energy (Bord, pg. 98) written in Sir Isaac Newton's *"Philosophiae Naturalis Principia Mathematica,"* Latin for "Mathematical Principles of Natural Philosophy"- one of the most influential scientific works ever published.

As earlier discussed, there must be a *medium* or substance that all of these different forces travel through. It is the means of which all wavelengths of the electromagnetic spectrum are transferred. It is the guide wave for the particle world; photons, electrons, neutrons, and protons are all influenced by its existence or exist, as waves of it and it may also be the pressure or force of attraction called gravity. If energy exists as waves in empty space, what is it that is waving? In the sea, it is water molecules, in the air, air molecules. All particles may influence or even be composed of this *medium*. As we saw before with the radio antenna, the motion of the electrons moving through the wire transferred energy into the *medium* and back from the *medium* into the receiving antenna even through "empty" space.

Gravity

What is gravity? What is it that causes the acceleration of objects as they fall? Galileo determined that all objects, not accounting for air resistance, fall at the same rate at 9.8 meters per second per second no matter what size or weight on the earth. Einstein noticed the similarity between gravity and acceleration and developed the "principle of equivalence", asserting that gravity and acceleration are the same. In a thought experiment, Einstein would imagine a man closed up in a in a "spacious chest" resembling a room. This room would be as if it were, in orbit with no gravity, when suddenly the room is accelerated by some force at a constant rate of 9.8 meters per second/ second. In this situation, not being able to see outside, the man perceives that this force is gravity, making gravity and acceleration indistinguishable (Einstein, pg. 67).

If gravity and acceleration are the same, how can we account for this "acceleration" in our universe? As a car takes off from a standstill and accelerates, the driver feels forced back into the seat. The higher rate of acceleration, the more the driver is pressed back. Now visualize the surface of the earth expanding outward at a rate of 9.8 meters per second, per second (per sec^2). All objects on the earth would maintain their weight similar to gravity. The problem with this however is that everything would also have to expand outward proportionally with the earth to stay in a balanced perspective and the rest of the universe likewise would have to expand at the same rate to stay in a proportional balance.

Now comes a dilemma; other objects with different gravitational forces must also be taken into consideration; for example, the moon, which has one-sixth, the gravitational force of the earth. With the earth's surface expanding outwardly at 9.8 meters per second per second, the moon's surface would be expanding only at 1.6 meters per second, per second (per sec2). It wouldn't take very long at all to notice from an earth perspective that the moon would be shrinking very rapidly. In order to correct for this dilemma one would again have to change space and time to straighten this dimensional warping of space. The rate of lunar expansion therefore may be perceived to have sped up in order to keep a universe in balance or perceived distances would have to be greater to have this perception of no change.

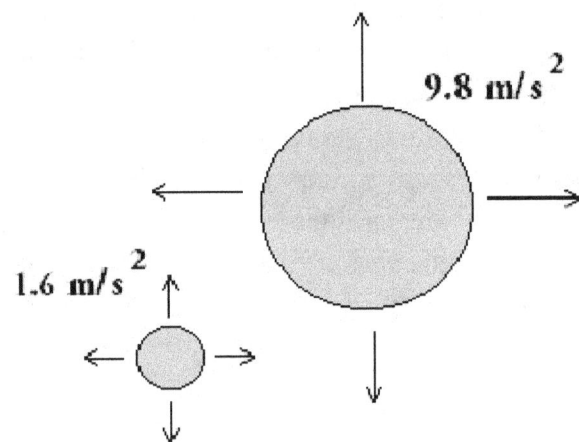

Different Rates of Expansion

This seems ridiculous. Again, we would have to resort back to the Lorentz Contraction Theory of dilated space and time in order to correct distorted space. However you look at it, all mass would eventually envelope all mass in this scenario, and everything would be annihilated; the earth would swallow up the moon and the sun, likewise the earth.

$$l' = l \sqrt{1 - \frac{v^2}{c^2}}$$

Dilation of space

$$t' = \frac{t}{\sqrt{1 - \frac{v^2}{c^2}}}$$

Dilation of time

Let's now look at another possible answer to this problem. If this gravitational force of acceleration is not caused by rapidly expanding mass, what else could it be?

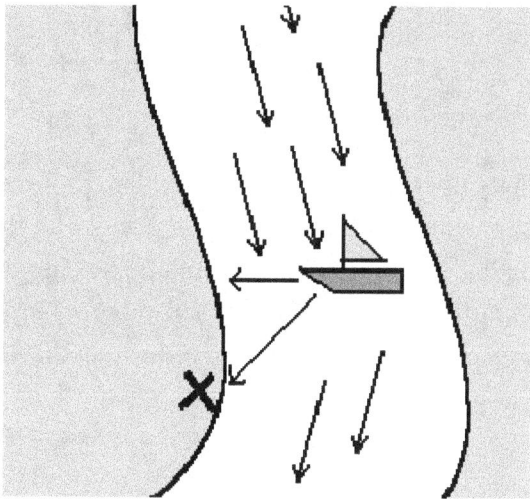

Flowing River

When a boat crosses a rapidly flowing river, the water current carries it along causing it to drift down stream. The same could be happening here with gravity. A rapid cooling of the universe after the big bang could be causing the condensation and rapid contraction of this medium, the aether cloud that surrounds all matter. As this material condenses, it forms in various levels of pressure, creating a "flow" of subatomic particles moving through and around all mass in its vicinity. This material would create regions of varying pressures; as the materials cool and condense the pressure drops causing higher pressures and energies to exist further away from the condensing mass. Higher pressures are outside of the condensing mass and lower pressures inside mass, this pulls in mass towards other mass, similar to wind on a sail.

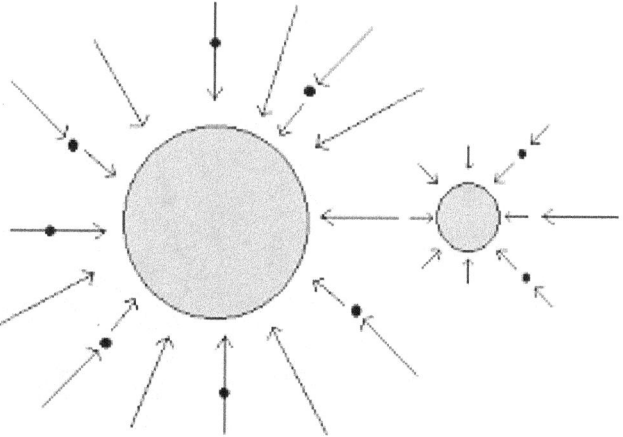

Condensing mass

The formula for the gravitational attraction of mass is:

$$F = \frac{M_1 M_2}{d^2}$$

It may be that subatomic particles, electrons, protons and neutrons are simply clouds of this condensing spinning energy in the aether. The rate of spin would be proportional to its energy and the direction of spin would be its charge. Like charges would repel each other like spinning tops bouncing off each other. Opposite charges would attract each other spinning in opposite directions moving together like gears in a massive efficient machine.

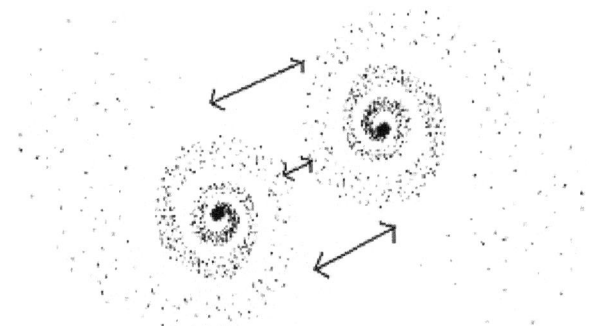

Like Spins Repel

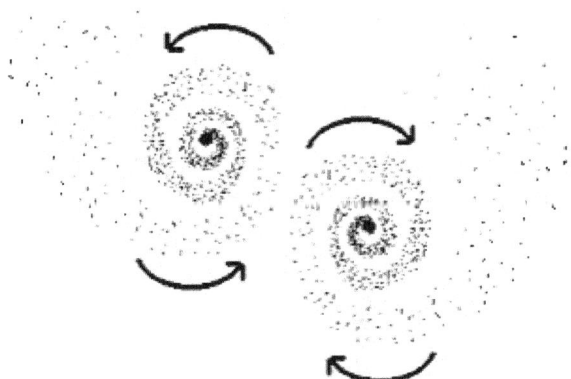

Opposite Spin Attract

Theoretically, it is these flowing and spinning clouds of aether energy, the same *medium* that transfers the waves of electromagnetic disturbances known as light, that cause gravity, electrostatic charge and magnetism. The particles of the aether do not necessarily have to flow; the "spin" may simply be energy waves trapped in orbit around these particles moving in different directions. These fields of "force at a distance" surround all mass and affect other mass in close proximity by its flow and spin without particles having to contact other mass. All of these forces decrease proportionally by the square of the distance from the mass-producing these fields. These fields easily radiate out into the open vacuum of space.

A boat crossing a river will move with the current, eventually moving at almost the same velocity of the water. The rate of acceleration depends on the boats mass and surface area to the force of the water. Unlike the force of the water on the side of the boat, the aether particles are so small that they exert a force on every particle throughout mass in general. Every atom, electron, proton, or neutron has a force on it from the pressures of the aether. Therefore since all mass is created of the same subatomic particles, all mass no matter what its density or size has the same gravitational force of 9.8 meters per second per second pressing on it here on earth.

A sailboat will accelerate as it is pushed by the wind. A sailboat however, can go as fast as three times the speed of the wind; how can this be? It is not the flow of the wind that moves the sailboat; it is the pressure differential on both sides of the sail that causes movement, in fact, a sailboat if sailed properly in a zigzag fashion, can sail into the wind. Likewise, it is not the flow of this aether that moves the object towards mass, it is the pressure differential between the higher pressures outside the central condensate particles and the lower pressures inside these central particles that cause mass to move together. The aether's differential pressure is proportional to the square of the distance to that mass, and the amount of mass. Differential pressure therefore increases proportionately closer to the attracting masses core an object comes. A large mass such as our sun will have a greater gravitational pull than our earth because the sun aether's has greater differential pressure.

Because of the earth's size, the majority of the aether flow (or pressure differential) moves through us and through all matter on the earth; every particle in every object is also absorbing this flow. Albert Michelson as we saw earlier designed a device he thought could measure the velocity of the earth as it moved through the aether – a device designed to detect aether flow, the Michelson Morley Interferometer.

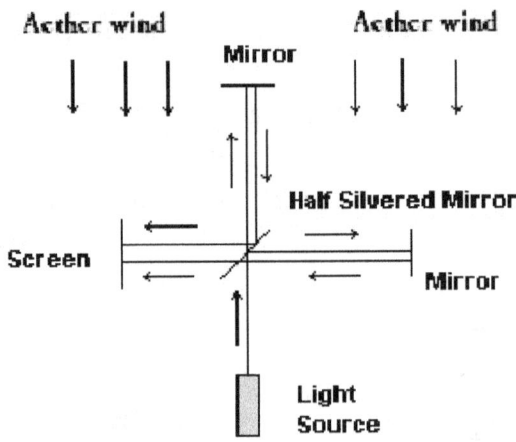

Michelson Morley Interferometer

As with the boat crossing the river, the current will drift the boat downstream, Michelson believed that an aether wind would drift light, causing his interferometer to display an interference pattern.

Interference Pattern

As the aether wind would flow across the light beam, it would slightly displace its reflection on the 45-degree half-silvered mirror. Half of the light would go through the mirror and reflect back again of another mirror. The aether wind would then slow down the light beamed into it. Both of these light beams would then be merged again at the silvered mirror toward an eyepiece. The slight change in distance that one light beam of light would travel (due to the aether wind) would cause an interference pattern in the eyepiece, as the light waves of both beams would now be out of phase. But this was not the case and the experiment failed. Why? Because they assumed that the aether flow was parallel to the earth's surface due to the earth's motion through space. In fact, this aether flow they were looking for may be perpendicular to the earth's surface flowing into it – known as gravity.

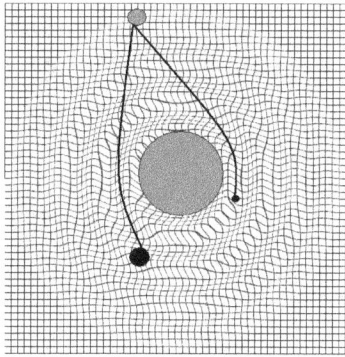

Radar Tracking of Mercury and Venus

Astronomer Irwin Shapiro's discovery in 1964 was in fact a modern day variation of the Michelson Morley Interferometer. Here, the beam was not light but a radar beam, and instead of reflecting off a mirror, the beam was reflected off the surface of the planets. Michelson and Morley were attempting to show that the light beam moving in the direction of the aether wind would be slowed down creating an interference pattern made by waves being out of phase. Shapiro showed that a shift in the radar frequency was created by radar energy being slowed down by the suns gravity. Here again, gravity is shown as the aether wind.

This is also evident during a solar eclipse when the corona of the sun can be seen around the shadow of the moon. This should not be the case. The size of the moon and its distance from the earth should block the suns corona totally. This was one of the first proofs that even the small gravitational field of the moon affects the movement of light around it. Starlight has also been known to bend or diffract by the suns gravitational pull, changing its perceived position in the sky. This was first seen in 1919 by astronomer Arthur Eddington while observing an eclipse off the coast of Africa (Isaacson, pg. 257).

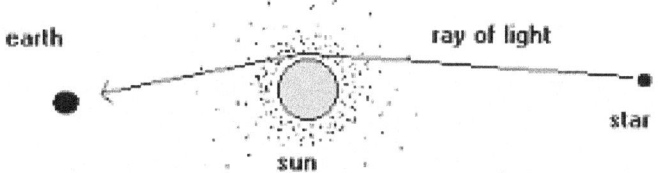

Diffraction of Light

At present, huge interferometers such as LIGO are being used to look for gravitational waves (Bartusiak, pg. 115). What are these gravitational waves? Could these waves flow through the aether, the substance that scientists are so reluctant to admit exists? Again, let's consider that all mass is slowly condensing material with the condensate material as the aether. As energy radiates out from the core, it cools and completes a phase change, turning it into an individual particle. As energy dissipates outward, the cooling material continues to condense, drawing more of this aether into it at a constant "flow". The closer to the core, the cooler this aether becomes, dropping in pressure similar to isobaric lines seen on weather maps depicting low pressure regions. High-pressure regions outside will then push material towards the low-pressure regions spiraling in a rotating motion.

This process over time has caused a cooling universe to condense or coalesce as mass attracts mass and starts to accumulate. If these different pressure regions did not generate during this process, the universe would stay a soup of thinly spaced particles, no light, just a dark cloud.

Over time, as mass attracts mass, large masses begin to accumulate more mass as their gravitational pull on smaller objects compounds dramatically. Now the large mass begins to heat up, squeezing out the energy it contains as if a sponge is being squeezed. Huge gravitational forces now cause the fusion of these fine particles into hydrogen and helium, the fuel sources for stars, and a star is born.

As the universe continues to cool, stars continue to grow until they run out of fuel. This fuel originally came from the dust and debris that the star pulled in through its strong gravitational field. Soon however, the space around the star cleared up, as if cleaned up by a gigantic vacuum cleaner and the star had to rely on the fuel it had stored within its self. Over billions and billions of years, this fuel slowly burned up, shedding its energy out into space, giving up much more energy than it absorbed. As its energy subsides, its gravitational force begins to overcome the energy supporting it from within. At first the star grows like a balloon in a desperate struggle to survive, but soon, with very little energy left to stay inflated can no longer support itself against its own gravitational field; it then contracts and collapses in on its self into a black hole. The gravitational force of its huge mass has overcome the nuclear fire which once raged within it, now nothing, not even light can escape this massive gravitational field. John Michell (1724-1793), English philosopher and author of *The Philosophical Transactions of the Royal Society of London* developed this theory back in 1783 (Hawking, 1988, pg. 83).

Not all stars come to this so-called end. It all depends on the mass of the star, very large stars however may indeed become black holes, and others become white dwarfs, brown dwarfs, neutron stars or pulsars. When medium sized stars like our sun collapse, they explode in a supernova, blowing off its gaseous material out into space and become neutron stars. More massive stars overcome by their own gravitational forces continue to collapse becoming black holes. Indian-born American astrophysicist and Nobel Prize winner Subrahmanyan Chandrasekhar (1910 - 1995) determined what is now known as the Chandrasekhar Limit, the threshold at where mass and energy levels allow for the development of a Black Hole (Hawking, 1988, pg. 85).

Black Hole's, once just a theory, are now popping up everywhere. It is believed now that every galaxy has one and may be what gave birth to the galaxies. As clouds of dust through gravitational forces pulled together, massive amounts of mass came together and developed mass beyond the Chandrasekhar Limit. Here mass levels were so high and gravitational forces so strong that energy levels within this congregating mass were too low to prevent it from collapsing and a black hole was born. Dust clouds in the close vicinity of the black hole started to spin at high speeds and the friction of the high velocity dust caused it to glow with enormous brightness, the creation of a quasar. The high energy and the enormous amount of mass being pulled towards the black hole generated the birth of stars, spinning at great velocities around the black hole. Dust and gasses near the center of this huge storm were slowly pulled into the black hole; the last light to escape glitters off the "event horizon", the last chance for energy to escape the massive gravitational pull of the black hole (Hawing, pg. 88).

Back to the scale of the stars, sun and planets this continues to be the case, causing mass to attract mass. The greater the mass is, the greater the attraction with a lower pressure; for example, between the sun and the earth; a higher pressure existing outside of both the earth and sun pushes them together. As light passes through this aether cloud surrounding a large mass, the density of the aether increases. The closer light comes to the large mass, the slower light travels, causing it to bend or refract, just as light bends while passing from air into water. The stronger this gravitational field, the denser the aether and the more light will bend.

Black holes may actually come in many sizes (Hawking, 1988, pg. 112). The Chandrasekhar Limit, although based on the amount of mass a non-rotating object can absorb before collapsing into a black hole, also must calculate the total energy of the object. For example, when a massive star runs out of fuel, at a specific energy level it can no longer support itself and will collapse. Mass is basically constant here and energy is the variable. Smaller stars may also become black holes in the depths of space where the stars mass can drop to temperatures close to zero Kelvin. Different amounts of mass therefore will become black holes at different temperatures.

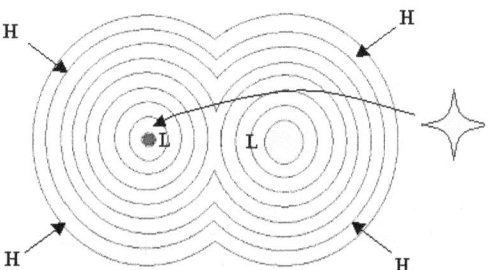

Isobaric Lines of Pressure

Although in today's scientific community, the concept of the aether is generally not accepted. Scientists however, do consider an invisible gravitational *field* to exist that surrounds all objects and increases in intensity with an increase in mass and decreases in intensity with distance away from the mass: the greater the gravitational intensity, the greater light is bent. What does this field of varying "pressures" consist of?

Magnetism

Magnetism is one manifestation of a field. An electromagnet is created with a coil of wire around a ferrous core material. When a battery is connected to the coil, an electron flow through the coil creates a field around the coils core called a magnetic field. The magnetic field contains lines of flux which seem to emanate from one end of the core, moving around to the other end, and can be seen by sprinkling iron filings on a piece of paper and putting a magnet under the paper.

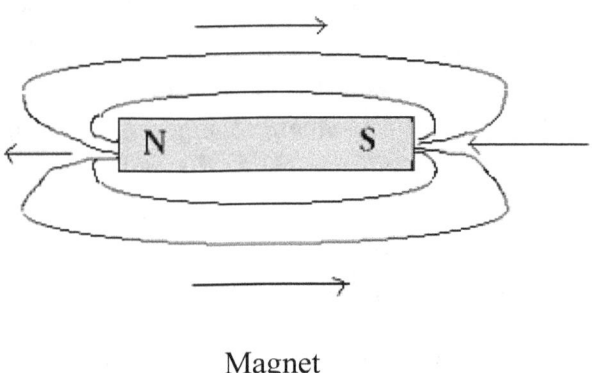

Magnet

What are these lines of force? In space or in a vacuum they exist but there supposedly is no mass associated with this flux. Like charges repel and opposites attract; when a north pole of a magnet is pushed toward another magnets north pole, the second magnet will move away. What *medium* transfers this energy?

As seen with the antenna, an alternating current induces a field, which flows with the current emanating out into space. With DC current in a wire however, current flows continuously from the negatively charged source of a battery towards the positive; with an electromagnet, the current flows from one end of a coil to the other. As the current flows through the coil, the field moves with the electrons, exiting one end of the core and returning back around to the opposite end outside the coil, then entering back into the core. This is the flux seen with the filings. In the next figure, the flux on the outside of the coil of both the south and north poles of the magnets flows in unison drawing them together. The flux flows out of the North Pole into the South Pole in unison with no opposition.

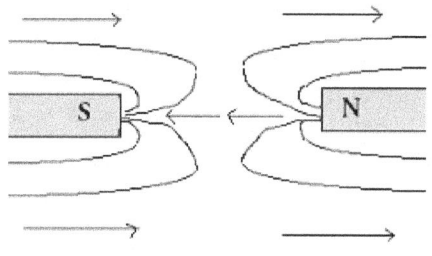

Attraction

The flow of flux out of the north pole of one magnet opposes the flow of flux out of the opposing magnets north pole. Likewise the flux flowing on the perimeter of the coil is in opposite directions. It is these opposing flux flows of the north poles that cause the magnets to repel. The same is true with two south poles of a pair of magnets. In natural magnets an electrical flow within the material itself possibly generates this field.

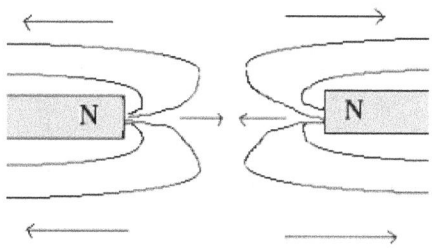

Repelling

Light Propagation Theory

What has been done here in the last few chapters of this book in order to account for the constant speed of light is to compare the propagation of electromagnetic energy to the propagation of sound energy. Since this supposed *medium* is a gas like substance dispersed throughout space, it moves and is compressed by moving objects. Energy then, as it enters into the *medium* from an object that is oscillating, moves through the medium at the speed of light in reference to the movement of the medium. Although the velocity of the source emitting the energy may vary, the velocity of light energy through the medium is constant. The velocity of the emitter however is directly proportionate to the wavelength of the energy received at the detector, and may be perceived as Red Shift or blue shift.

Picture two submarines in the depth of the sea sounding their sonar's in an effort to view obstacles. As Sub 1 is stationary in the water, it pings its surroundings with its 8,000 Hz sonar with the speed of sound in seawater at 3,355 mile per hour producing a wavelength of 0.1875 meters. The sub begins to move and speeds up to 25 knots (28 mph) through the water. As the sonar exits its transmitter into the water, each wave moves forward at the speed of sound in the water. The sub is also moving forward as it sends out the next wave slightly ahead of where it would be if the sub were stationary. As these waves are transmitted through the water, the subs forward velocity compresses the wavelength shorter down to say approximately 0.181 meters, closer together.

A second sub (Sub 2) arrives on the scene, and approaches Sub 1 at 25 knots. As the sound waves passing through the water enter Sub 2's sonar detector, they compress even more by the subs forward velocity through the water to say, 0.175 meters. Now, the speed of sound leaving Sub 1's transmitter was at the speed of sound to that transmitter. Once the sonar wave entered into the water, it transitioned to the speed of sound in the water and was in no way sped up through the water by the subs forward movement. As the sonar waves entered into Sub 2's detector, the sound transitioned from the water into the detector at the speed of sound referenced to the detector. The speed of sound transitioned through three different mediums with the final speed of the sound inside the detector, in no way being influenced by the movement of Sub 1, the water or Sub 2 towards or away from each other.

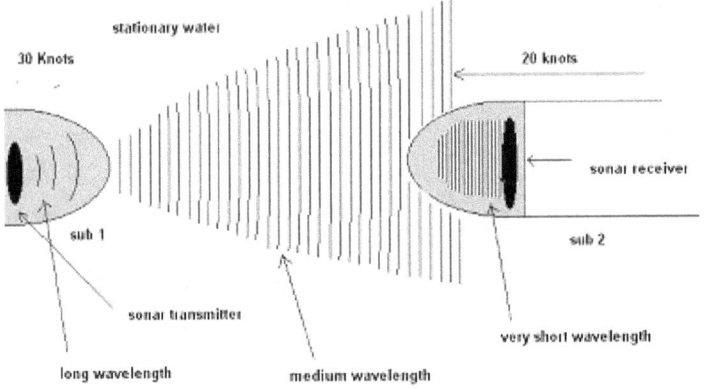

Sound Energy Propagation

In fact, there could be twenty subs all moving at different speeds through the water sending out signals for the other subs to detect. Each signal, as it transitions into the water from the sub transmitting the signal, moves at the speed of sound in the water independent of the subs speed, although the frequencies may vary. Each sub receiving the signal, likewise transitions the sound into its sensors from the water. The speed of sound inside the sensor is the speed of sound in that sensor and is the same for all subs no matter what their speed through the water may be. The frequency received however will vary by the speed of that sub according to the Doppler Effect.

The speed of light also is always a component of the medium it is traveling through. As light leaves a star in the great Andromeda Spiral Galaxy, it moves toward the earth at about 50 km/s (about 30 mi/s.), transitioning from the atmosphere of the star, out into open space. The light traveling through open space, similar to sound traveling through water is in no way sped up through space by the stars forward movement. As the light enters into the earth's atmosphere, it transitions into the medium moving with the earth and on into the medium in and around the detector. The detector then detects the lights velocity in reference to the medium it is in, not the earth's movement through space. The speed of light transitioned through three different mediums with the final speed of the light in no way influenced by the movement of the star, space or the velocity of the earth.

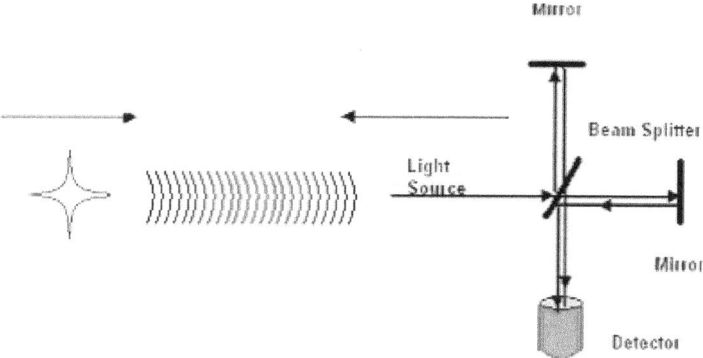

Light Energy Propagation

The speed of light in this context is measured the same no matter what the observers and detectors velocity is on the earth, what velocity the emitter or in this case, the star is, either moving towards or away from the earth and is consistent with Einstein's Theory of Relativity. The velocity of light however, just as with sound, is dependent on the medium that it is traveling through. The perceived wavelength however, is dependent and a direct indicator of the velocity of the object the energy left and the velocity of the object detecting it.

Sound vs. Light

It appears then that sound and light are very similar in propagation characteristics. Both seem to emanate in waves and propagate at fixed velocities in reference to the medium that their energy passes through. It also is apparent that the velocity of both sound and light is always the same to the detector no matter what the velocity of the emitter or the detector is in reference to the energy waves. Likewise, moving emitters or detectors generate the common phenomena called the "Doppler Effect."

What then would happen if we took the same mathematical equations developed with the propagation of light theories and applied them to sound wave propagation? Let's look at Einstein's basic formula and simply change the constant "c" to the constant "s" for sound.

$$E = MC^2$$

134

Energy therefore would be the mass of an object multiplied by the speed of sound squared.

20 kg X 331m/sec²

Or 2,191,220 kg m/sec

Theoretically we could increase the mass by adding energy since the speed of sound is constant; let's go with an energy level of 3,000,000 kg m/sec.

Or 27.4 kg of mass increase by adding energy

The next formula is very similar:

$$m = \frac{m_0}{\sqrt{1-\frac{v^2}{c^2}}}$$

With the speed of sound through air at 331 m/sec and a mass of 20 kilograms we calculate mass at half the speed of sound to be:

27390.25/109561=.75; sqrt of .75= .866; 20/.866 = 23.094 kg

Or an increase of 3.1 kg in mass

If we bumped it up to 90% of the speed of sound:

297.9/109561=0.0027; sqrt of .0027 = 0.0521; 20/0.0521 = 383.5 kg

Or a massive increase of 363.5 kg in mass

We could go on and work out these other formulas in the same matter, replacing the "c" with a speed of sound constant "s", with the assumption that nothing can go faster than the speed of sound.

$$t' = \frac{t}{\sqrt{1-\frac{v^2}{c^2}}} \qquad l' = l\sqrt{1-\frac{v^2}{c^2}}$$

$$x' = \frac{x-vt}{\sqrt{1-\frac{v^2}{c^2}}} \qquad t' = \frac{t-\frac{vx}{c^2}}{\sqrt{1-\frac{v^2}{c^2}}}$$

$$c' = c_1 + v\left(1-\frac{1}{n^2}\right)$$

Does this make sense?

Here or There

Zeno of Elea, from the pre Socrates Greek era (late 400's BC), wrote about a race that could not be won, a paradox of sorts. In the story "Achilles and the Tortoise" a tortoise challenges Achilles to a race, as long as Achilles gave him a small head start of only ten meters. The tortoise would then argue that as Achilles would start to catch up, the tortoise would cover a little more distance of which Achilles would also have to also catch up on. This would then go on indefinitely and Achilles could never win.

We can look at this scenario another way; let's start by going from point A to point Z where each step is broken into halfway points. To get to Z from A, you must go through B the first halfway point. Once at B, you must then go through C, the next halfway point. To get to Z you must yet go through D and so on. The points keep getting smaller but you are still not at Z. Theoretically, this can go on for infinity, because you can never get to an end point unless you first go through the next halfway point. Hypothetically then, the smallest divisible points will then be Y and Z. Y being here and Z being there. There is no existence other than these positions, no halfway point. These positions could be the particles of the _medium_. When they move from here to there they move almost instantaneously, not existing unless they are here or there (Wolf, pg.14).

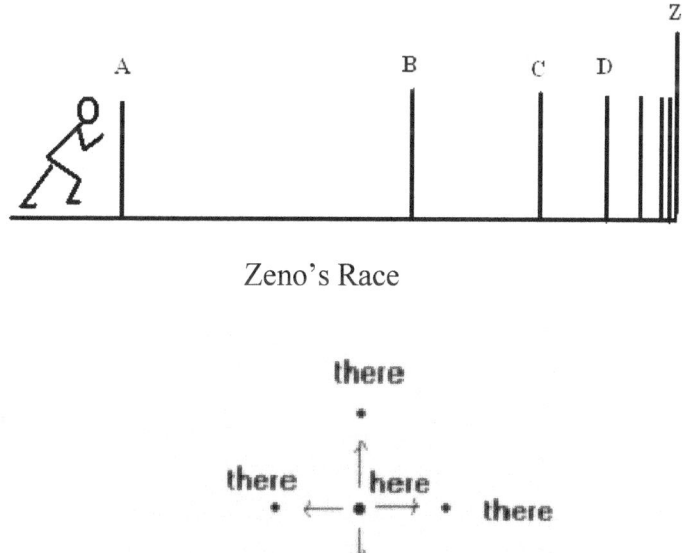

Zeno's Race

Here and There

In normal everyday life, we can measure say a car and see its location and its state of motion; this also goes for the earth around the sun and a bird in the sky. In the quantum world however, one can only know an objects position or its state of motion (Wolf, pg. 111). Because of the minuteness of this world, even the act of measurement changes the outcome of the measurement. A beam of light for example, being used to measure a particle, will cause the particle being measured to move or change. Therefore, only one measurement such as its velocity can be known, the other desired measurement must be calculated through probability. In order to mathematically calculate these motions and positions, a particle had to be portrayed as a wave function or quantum wave, according to the Copenhagen Interpretation of Quantum Mechanics (Wolf, pg. 261).

The act of observation then collapses this wave; probabilities must therefore be used to calculate the position of a particle without directly observing it. Werner Heisenberg pictured the quantum world as a fuzzy, spread out world, a world where specifics could not be known. He stated that the wave nature of particles makes it impossible to know the particles position and momentum. In fact, the more precisely you know a particles position, the less likely you are to know its momentum, only probabilities can be calculated; this is known as the Heisenberg Uncertainty Principle (Bord, pg. 388).

Here the very act of observation changes things; even the state of the universe ultimately could be at stake with a change in observation. In fact, according to the Copenhagen Interpretation, until an observation is made, a particles position and state are mere probabilities. Therefore, nothing would exist until it is observed. Einstein would go on to say, "The Heisenberg-Bohr tranquilizing philosophy — or religion? ...Is so delicately contrived that, for the time being, it provides a gentle pillow for the true believer from which he cannot very easily be aroused" (Isaacson, pg. 453). Einstein would ridicule this philosophy as "spiritualistic" and "spooky action at a distance".

Erwin Schrödinger along with Einstein would then look for another way to describe this interaction of particles in the "quantum world". Schrödinger then devised a way to describe particles with his famous wave equation. This equation would explain the probability of a particles position and state when observed (Brown, pg. 88).

Einstein did not really like this probability "stuff". He would describe this way of thinking as taking two boxes and placing one ball in one of the boxes; stating that the probability of the ball being in the first box as 50% chance was an incomplete statement to him. The right way to address this was to state that the ball was in the box or the ball was not in the box; that was reality! (Isaacson, pg. 455). Many would argue that the ball was in no means in one of the boxes until they would lift the covers and observed it; the "if a tree falls in the forest" scenario. This was that "religious like" new science that many believed in and Einstein was not so fond of.

Erwin Schrödinger would come up with another analogy similar to Einstein's box. "One can even set up quite ridiculous cases. A cat is penned up in a steel chamber, along with the following diabolical device (which must be secured against direct interference by the cat): in a Geiger counter there is a tiny bit of radioactive substance, so small that perhaps in the course of one hour one of the atoms decays, but also, with equal probability, perhaps none; if it happens, the counter tube discharges and through a relay releases a hammer which shatters a small flask of hydro-cyanic acid. If one has left this entire system to itself for an hour, one would say that the cat still lives if meanwhile no atom has decayed. The first atomic decay would have poisoned it. The psi function for the entire system would express this by having in it the living and the dead cat (pardon the expression) mixed or smeared out in equal parts". It is typical of these cases that an indeterminacy originally restricted to the atomic domain becomes transformed into macroscopic indeterminacy, which can then be *resolved* by direct observation. That prevents us from so naively accepting as valid a "blurred model" for representing reality. In itself it would not embody anything unclear or contradictory. There is a difference between a shaky or out-of-focus photograph and a snapshot of clouds and fog banks" (Isaacson, pg. 456).

Einstein loved it: he would go on to say "Your cat shows that we are in complete agreement… a psi function (wave function – also shown as ψ) that contains the living as well as the dead cat just cannot be taken as a description of a real state of affairs" (Isaacson, pg. 457).

Time and Space

When it became obvious to scientist that the speed of light was not only constant, but also that light was always detected at the same speed no matter what the velocity of the source or detector, scientists became perplexed. Albert Einstein would go on to say that this dilemma had "plunged the conscientiously thoughtful physicist into the greatest of intellectual difficulties" (Einstein, pg. 18) and came up with a brilliant solution that would remedy all – warp time and space under the effects of gravity and various velocities.

There is another possibility however, that we have been subtly suggesting throughout this book, that may be more reasonable and may work just as well scientifically, another way of altering space and time with the existence of the aether. But didn't the Michelson Morley Interferometer disprove its existence? Doesn't research on Binary stars, gravitational red shift, Mercury's precession and GPS satellite clock anomalies prove that time and space warps with high velocities and strong gravitational forces?

Let's perform another thought experiment where two spacecraft are traveling at two different velocities relative to the same starlight. One spacecraft is traveling towards a star at 150,000 kilometers /second or half the speed of light relative to the star. Another spacecraft is traveling parallel to the first spacecraft at 75,000 kilometers / second or one quarter the speed of light relative to the same star. Both of these spacecraft have observers with test equipment measuring the same light but are traveling at different velocities relative to the star, yet both measure the same light at 300,00 kilometers / second.

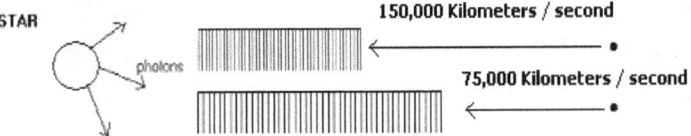

Relative Contraction of Space and Time

What we seem to see here is that the space and time between each spaceship and the star is being compressed and that this compression or contraction of space occurs according to the "Lorentz Contraction" of space relative to each of the spaceships velocity. According to Einstein however, each spacecraft would have a different rate of time passage and different measurements on board due to their individual velocities relative to the star; it is not space between the star and the spaceship that is altered, it is the ship itself.

Now, if one spaceship would come between the star and the other spaceship blocking the starlight, suddenly the focus is different. The light the trailing space ship then is viewing is the light reflected off the spaceship in front of them traveling at a slower speed, rather than the light of the star. Did the space ship suddenly shift its shape to adjust to the light reflected off the ship in front of it? If this were true, the ship would have to reshape according to all objects giving off light and moving at different velocities relative to it, compressing for some and expanding for others. Similarly, for binary stars moving in opposite directions to an observer, would both stars stretch and compress at the same time to accommodate for the speed of light to different observers viewing them in different regions of space? Of course not.

$$y' = y$$
$$z' = z$$

$$x' = \frac{x - v\,t}{\sqrt{1 - \frac{v^2}{c^2}}} \qquad t' = \frac{t - \frac{v x}{c^2}}{\sqrt{1 - \frac{v^2}{c^2}}}$$

Lorentz Contraction

The velocity of the light detected from an object can only then be relative to the observers themselves; the observer's actions cannot affect the motions or observations of other observers and the space-time around them. Only viewing velocity, time and space in this matter prevents a paradox. This also allows for any velocity to occur without changing space and time; therefore an observer's time and space cannot change with their velocity as previously thought. If time and space is not being warped by the movement of an object, how then is the speed of light the same velocity to all observers, no matter what their velocity or the emitter's velocity?

This brings us back to the submarine illustration. The received sonar signals speed inside the sonar dome of the receiver is at the speed of sound and is constant inside that dome, no matter what the speed of the sub emitting the original signal, or the speed of the sub receiving the signal. The velocity of the water currents between the two subs also has no effect on the sounds velocity inside the sonar dome after it is received. Likewise, the velocity of sound coming down the ear canal of an observer listening to the trains whistle also is unaffected by the trains velocity or the observers own velocity. Sound always has the same velocity; it is only changed by the temperature and pressure of the air it is traveling through. The only thing that can change due to the movement of objects then is the frequency of the detected sound, the Doppler Effect as previously reviewed.

The same must be true with light. The velocity of light inside the detector, whether it was Bradley's telescope, Fizeau's and Foucault's instruments, or the Michelson Morley Interferometer will always be the same. This velocity will always be 186,000 miles per second or 300,000 kilometers per second, no matter what the speed of the emitter (train), movement of space between the emitter and receiver (wind) or the movement of the receiver itself (observer). This light moves in waves within some rigid medium and will only vary in velocity depending upon the density of that medium; only the frequency of the light received will vary according to the movement of the various elements, as seen with the Doppler effect as red or blue shift. The two spaceship previously discussed will also have the same process occur. It is not the spaceship that contracts nor is it the space between the star and the ship that contracts, it is the light waves themselves that change as they move from one medium to another. The speed of light inside the detector remains the same, no matter the movement of the ship or the star – just like the submarine.

Time and space here are still affected. The speed of light will vary according to the type of medium it travels through, for example light traveling through glass as compared to light traveling through air. The time it will take light to travel a specific distance will then vary; space also will then vary proportionately with density. Space then also becomes denser close to mass through the cooling process or phase shift as previously discussed and less dense the further away from mass.

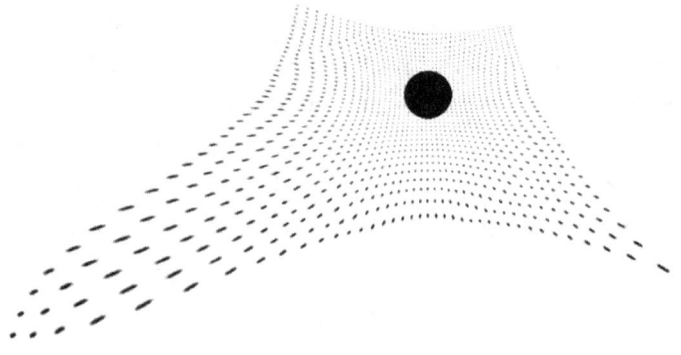

Warped Space

Einstein made an analogy using a bowling ball on a trampoline to describe how space warped with large masses such as our sun. The bowling ball would cause the rubber sheet of the trampoline to sink in the middle, as objects such as billiard balls were rolled onto the rubber sheet; the balls would roll inward toward the bowling ball in a spiral manner (Isaacson, pg. 223). This analogy was Euclidean or two dimensional in description, in real space-time there would be an infinite number of sheets surrounding this ball in three-dimensional space with many layers. In three-dimensional space, it would be spherical sheets much like that of the layers of an onion. Each of the layers of onion would represent a change in gravitational field strength, a change in field strength density if you will. The closer to the bowling ball, the tighter together the sheets become and likewise, the further away the more distance between the rubber sheets or onion layers.

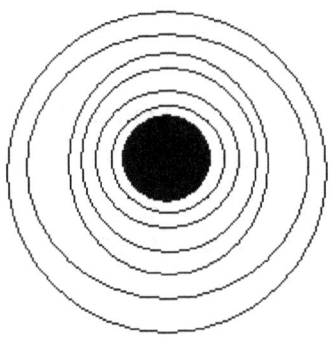

Onion Analogy

What is this field made up of which changes in strength? What would change in density? As we reviewed waves in previous chapters we also asked the question, "What was waving?" Could it be the same material?

In the previous chapter "Here and There," we reviewed that if we were to keep dividing distances in half, which eventually we would come to a point where division was no longer possible; a particle was here and another particle was there, nothing was in between. We can look at these particles as the field that exists in "empty" space, the medium that transfers energy across great distances, and whose condensation creates elementary particles. Could it be the variations in the density of these particles that cause light to bend and the differences in energy levels that create the pressure differential that causes gravity? This could be the onion we are looking for.

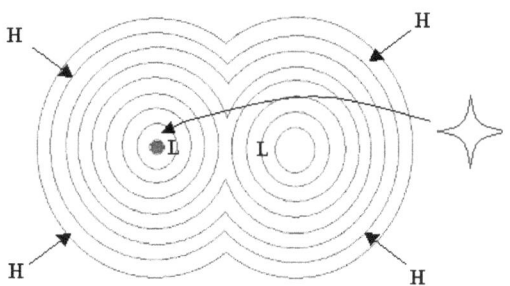

Warping of Space

An example of this is the speed of sound in water. At 100o Celsius (boiling), the speed of sound in water is 1,543 meters per second; however at 0o Celsius (freezing) the speed of sound is 1,402 meters per second. The closer together or the denser these particles are, the longer it takes for energy to pass through them; this is what causes light to bend around large objects in space. The closer to the earth, sun or any other condensing mass such as subatomic particles, the denser the medium becomes that light waves move through; the denser the medium the slower the energy moves.

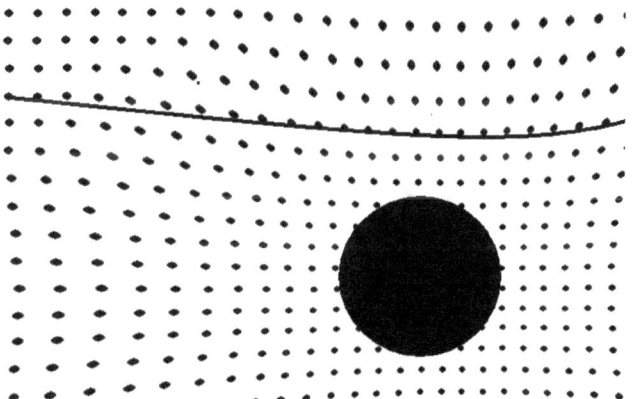

Warped Space Denser Closer to Massive Objects

If mass is merely created out of standing waves within this medium, then the velocity of these waves will also slow down as the medium becomes denser. Motion will be dampened, clocks will tick slower and redshift will be noticeable in light waves emanating out of these objects with large gravitational fields - in fact time will slow down inside of these gravitational fields. This gravitational effect has been clearly noticed in the timing circuits of the GPS satellite system covered earlier.

The larger the mass, the slower the time; likewise the deeper into these onion layers an object moves, the slower the time. Super massive objects such as black holes may indeed slow time down so much in the center that time stops and a singularity similar to the one that existed at the beginning of the "big bang" could exist. The difference here is that this is a miniature "big crunch" or a collapsing of matter, rather the massive expansion of matter during the "big bang." If time goes slower the stronger the gravitational fields, then the reverse should be true outside the gravitational effects of large masses – time should then speed up (Hawking, 1988, pg. 90).

The "Theory of Relativity" also states that velocity also changes time. In the section "Blast Off", we reviewed relativity and how that a moving object's velocity is only relative to the position of another object. For example, if a spaceship was far out in the universe with no object to truly reference velocity by, and came across two objects such as asteroids moving away from each other, what would be used to reference the speed of these objects? If they were floating apart at 500 meters per second, was object "A" moving at that speed away from a stationary object "B"? Or was "B" moving away from stationary "A"?

Space Objects in Motion

If "A" was moving faster than stationary "B", time would be slower on "A" but if "A" was stationary and "B" was moving away then "B's" time would be slower. Velocity here is relative and therefore velocity can have no effect on time on the moving objects. There is one catch however, if both objects were stationary relative to each other at one time and one began to move, the object that started to move away had to accelerate. The force of acceleration in the "Equivalence Principle" is the same as gravity – they are indistinguishable. Therefore, it is the acceleration of a moving object that must give it its time change, not velocity itself. Unless a single point of reference for all objects is determined, velocity is simply relative. Only when we have determined that this solid medium that energy waves are moving through exists, then do we have a reference to measure the speed of light to – a stationary point in space.

Velocity, no, rather a change in velocity, well, let's say a constant change in velocity due to acceleration will then slow down time. The longer and the stronger the acceleration forces, the slower time becomes. Therefore, if an astronaut launches out into space and accelerates up to great velocities, time on earth will pass faster than time on the spacecraft. When he arrives back, more time will have passed on earth than in the space ship – he would be younger than if he had never left. If an astronaut has to accelerate leaving the earth, he will likewise also have to decelerate to slow down and return. This deceleration as we saw before, simply changes the direction of the force from aft to forward. The effects of deceleration are identical to acceleration – they therefore are identical. As long as this force exists, time will slow down. Why leave the earth then, why not use a spinning acceleration machine instead to change time? The amount of high forces due to acceleration or gravity may change time but it may be so negligible that it would never be noticed.

We have seen that time may be slowed down due to the effects of gravity or acceleration. Can time be reversed though? Can we go back in time? In his book, A Brief History of Time, Stephen Hawking suggests that even though the Theory of Relativity states that speeds faster than light are impossible, if one could do so, he probably could return before he left (Hawking, 1988, pg. 163). Simple math however prevents this. No matter how much you multiply a velocity by, you can never get to zero time. For Example, at 100 miles an hour, it will take you one hour to reach the one hundred mile mark. You multiply your velocity by two; you will reach the one hundred mile mark in a half an hour. Multiply by three, one third of an hour and so on. No matter how fast you go, you cannot go faster than time.

Time is not the only thing that changes; space also is changing (which in effect changes time). We studied the possibility of the "Big Bang" how that the universe and all that we know originated out of a singularity, a point where time and space were "infinite". A huge blast then occurred, when the natural balances that contained the singularity broke down, blasting matter and energy out into the far reaches of the universe. Energy, not being able to exist without mass (it is simply the movement of mass) must be associated with some sort of particle, the particle that all mass consists of, the "God Particle". These particles originally were spread out evenly throughout the universe, but as energy was passed on from one particle to another, this energy was dispersed out and away from the center of the blast. Higher energies to the outside of the blast area forced cooler particles in the center of this area to coalesce and bond together, a phase transition of some sort. This was the foundation blocks of the stars and galaxies that we see today.

As energy slowly is dispersed out and away from the original blast sight, the center continues to cool and contract. Stars and galaxies further away from the center of this blast should therefore have more energy, expanding faster than the cooler areas. Over time, the whole universe should be also cooling uniformly (although areas further away are hotter). Areas further away however, also take longer to view from earth so that objects further out are seen at higher temperatures and at velocities than what they may be today; these areas have since cooled down and slowed down, possibly dramatically.

Edwin Hubble's discovery in 1929 suggested that galactic Red Shift in light from stars and galaxies increase in proportion to their distance from earth. His studies showed that 38 of 40 galaxies observed are rapidly moving away from each other, an omni-directional galactic expansion where galaxies furthest away are moving the fastest, some at 57,000 kilometers per second or 13 million miles per hour or 2% the speed of light. The problem here is that this is evidently omni-directional, indicating that we are the center of the universe. If we were viewing this expansion out on the edge of the universe, only certain areas focused in one region would have the greatest red shift. Either we are the center of the universe, or this observed redshift is simply an illusion.

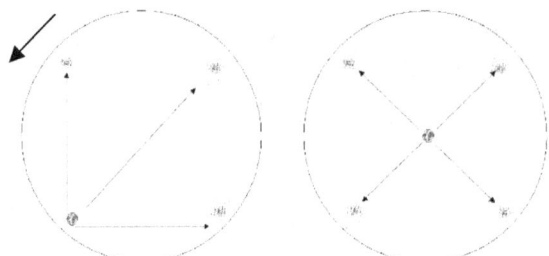

Since it takes time for the universe to cool and also takes time for light to reach us on earth, the changes to the universe will be isotropic or seen uniformly in all directions and be relatively consistent at the same distances. This however suggests a possible "Big Crunch" as energies subside and particles coalesce. Eventually as the medium between all objects condenses and cools, there will be nothing to keep objects apart. Higher energies outside will eventually push all matter together into one place. Here atoms can no longer exist, fusing together into what has been called the **Bose–Einstein Condensate** or BEC's, huge molecules that form at near 0 Kelvin or **absolute** *zero* – cold black holes if you will (Pitaeskii, pg. 1).

Energy however, would have to go somewhere; it cannot just disappear according to the **Principle of the Conservation of Energy**. If space is infinite in all directions, over time all energy would be evenly dispersed omni directionally and the universe would eventually grow cold and die. Now, if we take the stand that all other matter, including ourselves, were merely energy waves (standing waves) moving through a stationary substance that varied in density, our existence would be limited. The further out in the universe we would go, from the center of the Big Bang, the less dense this substance would be, to the point where matter could no longer exist; much like how sound cannot propagate and exist out in space. Energy cannot move through nothing, therefore energy would have to slowly migrate back into the known universe, reflected back in from the edge of time and space, pressing in on mass in the form of gravity.

If energy however was forced to loop back in at the edge of the universe or loop back around though **black holes** into other universes that constantly feed each other, then it is possible that matter and energy may be able to continue on indefinitely. Many different theories have been developed over the last fifty years trying to show how it all works, from a universe in the shape of a doughnut where time and space loops back in on itself, to the **"String Theory"** of multiple universes. If we think about it, a black hole's singularity draws in matter; the singularity that occurred at the **"big bang"** gave off matter. Can one singularity be the source of the other singularities energy? Is this the entrance to a parallel universe?

Cosmology

Orbital velocities of bodies, in our solar system, normally vary according to Keplerian Dynamics; orbital velocities that are proportionate to the distance from the object they orbit. The further away from that object, the slower the orbital velocity of the planet or moon, decreasing this velocity inversely with the square root of the orbital radius from that object, this is due to lower gravitational forces. For example, Mercury takes 88 earth days to orbit the sun while the earth takes 365 days. Mercury's orbital velocity is 48 km/s while the earth's orbital velocity is slower, about 30 km/s in it voyage around the sun. Likewise, Neptune's orbital velocity is only 5.4 km/s and it takes about 164 earth years to orbit the sun.

This, however, is not the case with stars orbiting in a galaxy. Surprisingly, most stars orbit a galaxy at approximately the same speed, in unison, as if tied together; stars on the outer fringes are therefore are orbiting much faster than they should, some at a radial velocity as high as 180 km/s, and theoretically should be slung off into space. As with interactions within our own solar system, orbiting objects are not only attracted to the object they are orbiting; they also have gravitational forces acting on them from all other objects relatively close to them. The earth is, not only attracted to the sun, but all other planets, including our own moon, causing the orbital distance from the sun to vary. Stars likewise, are not only attracted to and orbit around a black hole in the center of the galaxy, they also are gravitationally attracted to all the other stars, planets and gasses moving around the black hole, as if in a soup, moving together. Calculations indicate that there must also be a missing mass that can't be seen in order to account for the necessary increase in gravitational forces needed to prevent these stars from flying off, this is the foundation for the **Dark Matter** theory.

Further calculations had to be formulated to account for additional missing matter, as only 25% of the universes matter could be accounted for. This was the foundation for the **Dark Energy** theory; the energy that helps accelerate stars and galaxies in the far reaches of the universe, faster away from each other, as manifested in Hubble's Law. This law states that velocity, at which galaxies are moving away from the Earth, is proportional to their distance from the earth. The cosmological formula most recently calculated for the universes total mass is 4 percent baryonic matter (composed of atoms and subatomic particles of some sort), 21 percent Dark Matter, and 75 percent Dark Energy. How does this work then? What are Dark Matter and Dark Energy? Let's review the Big Bang.

As discussed previously, the Big Bang theoretically occurred some 10 to 20 billion years ago when something happened within a singularity and thermonuclear energy within the singularity could no longer be contained. A huge amount of energy was released and the smallest of particles blasted out into space at enormous velocities, dissipating their energies out into the blackness of the universe. As this outward energy dissipated, a reversal of the flow or an "implosion" of these particles of mass begins to take place, back toward the center of the blast and towards each other, slowly condensing these smallest of particles into baryonic matter. The outer regions of the universe now contain much of the energy released in this blast as the origin, or center of the blasts begins to cool. At some point in time, the energy blasted out into the universe greatly diminishes and dissipates as the dispersion of energy decreases by the square of the distance it travels away from its source.

The question now being asked is, 'are particles and objects closer to the Big Bangs source moving back towards the origin of the blast, as if sinking into a Black Hole?' If this is the case, could these objects (galaxies, stars, and even us) be "moving away" from the galaxies at the fringes of our universe, rather than those galaxies "moving away" from us? Theoretically, the further away from the origin of the blast an object would be, the weaker the pull on objects towards the center of the Big Bang, giving the illusion of variance in velocity in relationship to distance. As matter condenses and becomes more compact, time also slows down, giving the illusion that galaxies seen at the fringes of our universe (looking back into time) are moving away much faster than they currently are in reality. Could it be that the Universe is no longer expanding and that Hubble's Law is an illusion; an illusion created by an accelerating collapse of "space and time"? The Big Crunch may be underway!

If Hubble's Law were simply an illusion, Dark Energy, the substance that supposedly accelerates galaxies at the edge of our universe, would no longer be required. All that would be needed would be Dark Matter and baryonic matter, materials consisting of atoms and subatomic particles. Dark Matter possibly could be the substance that all baryonic matter is made of, also known as the aether, what all energy may move through.

Hubble's Law – The further away, the faster objects (galaxies) move away from the earth, causing red shift. The earth would have to be at or close to, the center where the Big Bang occurred.

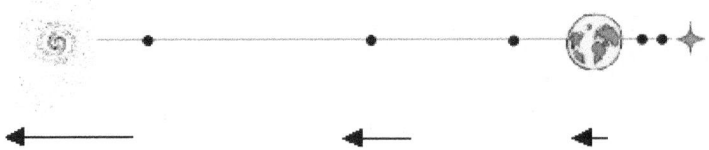

Inverse of Hubble's Law – For mass closer to the start of the Big Bang, the faster mass moves towards the center (as if falling into a Black Hole) creating Red Shift and causing the illusion that further away objects are moving faster away from us.

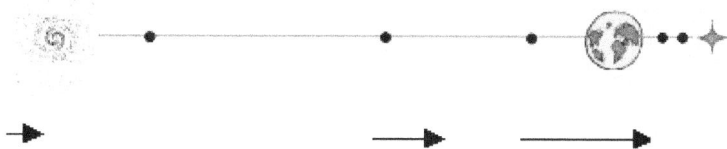

Triangulation has been used for centuries to calculate distances or the height of terrestrial objects, such as trees or mountains. A known length, or height, at a known distance, is compared with an unknown dimension at a greater distance, in order to calculate an approximate measurement; a pencil at arm's length compared to a tree for example. One observable variance similar to triangulation with stellar objects is called parallax. Parallax occurs when our viewpoint relative to nearby and distant objects changes, creating an apparent angular change in position of the nearby objects, relative the distant objects. This effect diminishes with distance so that even for the stars nearest to us, the parallax is so small that extremely precise measurements are needed just to detect the effect. Due to the earth's rotation around the sun, the distances to planets such as Jupiter (although it also moves in orbit around the sun) can be approximately calculated. Parallax can be observed as the earth moves in orbit around the sun over time; here stars closest to us in our own Milky Way galaxy seem to wander slightly, when compared to the stars around them that are further away.

However, this process cannot be used to calculate the distance of distant stars and galaxies. Astronomers must use luminosity to calculate the distance to an object, such as a galaxy, by measuring the brightness of the object and applying the inverse square law. The problem is stars vary in size, color and brightness; even galaxies vary in these attributes and are almost impossible to calculate unless you find a standard brightness or candle to go by.

The best standard candle for determining the distance to the nearby galaxies is the Cepheid variable star;, a star that has a precise level of luminosity and is very common throughout the universe. Some of these stars are even close enough, in our own galaxy, to measure their distance by parallax. With a known distance and a known luminosity, the distances to other Cepheid variable stars in other galaxies can now be calculated. Edwin Hubble used Cepheid variables stars that he recognized in the Andromeda galaxy (our closest neighbor), to approximately calculate its distance from us at 2.5 million light-years. While cataloging the distance to these various galaxies, Hubble discovered that the light signatures from the furthest galaxies away from us had the most red shift; varying proportionately with their distance from us, thereby establishing Hubble's Law.

String Theory

String Theory is one of the latest attempts through very detailed mathematical calculations to explain quantum mechanics using vibrations of one-dimensional strings, which only have length. These strings are the constitutes of all particles and are the foundations of time and space. Various vibrations at different harmonics, such as the strings on a violin that vary in pitch and amplitude, can represent various components of subatomic particles such as quarks. Different combinations of these strings vibrating at different frequencies manifest themselves as electrons, protons and neutrons. Combinations of these strings are called membranes or **"branes"** for short. Different branes may represent different universes and the collision of these branes, like sheets hanging on a clothesline bumping into each other, may be the source of the "Big Bang". Strings can be open or closed; one closed string manifests itself as a graviton, the source of all gravity (Hawking, 1988, pg. 176).

Open and Closed Strings

Similar to Theodor Kalusa's five dimensional theory proposed in 1919 (Isaacson, pg. 337), M-Theory envisioned in the late 1990's, is one of the latest string type theories that developed 11 different dimensions, 10 spatial and one time: M standing for membrane. Super String Theory is the latest attempt to combine and synchronize the various string theories (five seemingly different theories) into one. It now seems that all of these theories were all explaining the same thing in different ways, all being supposedly correct. All of this was developed with huge mathematical formulas based on the existence of a one-dimensional and two-dimensional world. Equations in the three and four-dimensional worlds became much more complex, explaining the eleven dimensions even more so (Ashtekar, pg. 316).

There is a very strong similarity between what has been suggested in this book and the **String Theory**. We have looked at the possibility that all that exists simply exist as energy waves manifested as **standing waves** moving through some semisolid substance. Solid because as far as we know, transverse waves of any kind can only pass through solids and cannot pass through fluids; light waves are clearly transmitted as transverse waves having polarity. Instead of minute particles oscillating at various frequencies and amplitudes as energy moves through this substance, the String Theory suggests that small one dimensional strings of various length, both open and closed in structure, vibrate at various frequencies and amplitudes to create the same effect –our world as we know it – a coincidence?

Unified Theory

Until recently, scientific laws have basically obeyed Newton's laws. Scientific discoveries of the 20th century however have uncovered a new science, one that does not seem to obey the same laws. As these different theories evolved, we have slowly developed one sort of law structure for the large and another for the very small. These two theories are somewhat different, as Tim Allen in his book, "I'm Not Really Here" states that the rules, "at a subatomic or quantum level are, well…. different" (Allen, pg. 10).

In his later years, Einstein struggled to make the quantum world and relativity compatible, a struggle to mathematically unify the gravitational and electromagnetic field theories into on **Unified Theory** (Isaacson, pg. 337). "It has been my greatest ambition to resolve the duality of natural laws into unity", Einstein said, "The purpose of my work is to further this simplification and particularly to reduce to one formula the explanation of the field of gravity and of the field of electromagnetism"(Isaacson, pg. 342).

Einstein struggle to his deathbed trying to find the answer for the Unified Theory, writing paper after paper, trying to solve the riddle; but page after page seemed to be flawed, losing the "physical" explanations of his older writings, moving towards "pure mathematics." Scientists of that day began to lose their passion and trust in his writings, and Einstein began to lose his touch, being rejected by many of his colleagues and became skeptical about his work.

Various forces are apparent in the universe, as we have discovered. These are electric fields, magnetic fields, electromagnetic fields, the strong and weak nuclear forces and gravity. All of these forces seem to hold both positive and negative characteristics, pushing and pulling, except for gravity. Combining these two theories into one became a nightmare! They just didn't work together.

Classical mechanics and Einstein's Theory of Relativity, although there is some things that conflict, work together much better than the theories of classical mechanics and quantum mechanics; they just did not work together! The **Heisenberg Uncertainty Principle** for example would not work well with classical mechanics. This is where the more precisely you know a particles position, the less likely you are to know its momentum, only probabilities must be calculated.

As we look back at the ideas generated in this book, we see that even through the simplicity of classical mechanics, gravity can be created. Going back to the "Big Bang" and the enormous energies that were released, to the cooling processes that caused mass to coalesce from the building blocks of nature (possibly strings), and the high energies outside of mass that draws particles together.

We also saw how as objects and particles spin, that particles within "empty" space between these particles also may spin (or at least the energy moving within these particles) affecting other objects at a distance. Opposite spins allow objects to move together and not repel while objects of like spins repel each other.

Building this type of universe where laws work together and that both small and large, the regular world and the quantum world move together is not out of the question.

Worm Holes

We have developed a hypothesis that a medium exists in which energy wave's move through; a medium that varies in density according to its "temperature" or "pressure". As this medium cools, it condenses, forming particles that are the foundation blocks of all matter; the closer to matter the medium is, the cooler and denser this medium becomes, in effect changing time and space. We have also hypothesized that matter may also be created by "standing waves" of energy passing through this "elastic-solid" medium. Close to the condensing mass, these energy waves move slower and tend to bend as they pass large bodies of mass, refracting if you will, distorting space and time. The closer to the large mass these standing waves come, the slower they move. This difference in density, creates a pressure differential which in turn causes the gravitational pull; the greater the mass, the greater this pressure differential and gravitational force.

The opposite may be true as these waves move out and away from large masses such as stars and galaxies. Just as light velocity increases as it moves out of the medium of a piece of glass in to air for example, energy moving through "near space" at the velocity of "c" may move much faster far away from large bodies of mass. The velocity of light in space has been assumed to be the constant "c", 186,000 miles per second. However, if space is not empty as we have believed, and light waves move through a medium that changes in density the closer to mass the medium is, and then the speed of light will also change according to its density of this medium it passes through.

If we picture space as a large ocean full of islands, we may also envision an ocean full of currents. As ships on the high seas use these currents today, we may likewise also be able to use the currents in outer space. Areas of space between stars or between galaxies may contain regions where space is almost void of this medium, allowing energy or mass in the form of energy (standing waves) to pass from one region to another at very high speeds – in effect, passing through a natural worm hole.

These wormholes may also possibly be generated mechanically. High beams of energy may be able to temporarily create a void in space, similar to how a lightning bolt superheats and ionizes air momentarily, followed by the sound of thunder. The momentary opening of this tube, mostly void of the medium would allow energy to move through it at speeds much higher than the accepted speed of light of "c", allowing us to "jump" from one region of the universe to another.

This "jump" may also be used for short distances, allowing for transportation from one area on earth to another instantly, "beaming" people or objects from place to place. Energizing or "heating" the medium in close vicinity to an object would decrease the mediums "pressure" on one side of the object. This pressure differential would cause a "gravitational pull" if you will, in a specific direction, moving the object without G forces, as if falling towards the targeted area. If the medium were energized below an object, it would cause the object to rise and be weightless. An object used as a transportation device, could be moved in any direction, and at almost any speed, without the adverse effects of G forces and acceleration /deceleration.

How to energize the medium or aether is the question. Specific wavelengths of electromagnetic energy may be the answer, similar to how water molecules are energized in a microwave oven.

Black Holes

The current Webster's Dictionary definition of a Black Hole is "a celestial object that has a gravitational field so strong that light cannot escape it and believed to be created due to the collapse of a very massive star". The scientific theory of a Black Hole is relatively new. As seen previously, geologist John Michell wrote about the idea over two hundred years ago in his 1783 paper, "Philosophical Transactions of the Royal Society of London". Here, Michell suggested that, if a star were to grow to sufficient mass, at some point, its gravitational force would be great enough to produce so strong of a gravitational field that even light could not escape from it. It wasn't until 1969 that the term was coined when scientist John Wheeler used it in a speech, and stuck (Hawking, 1988, pg. 84).

Soon, new ideas began to surface and Black Holes started to become more scientifically complex. The core of the collapsed star now had a "singularity" similar origin of the Big Bang, and the outer shell of the Black Hole was now called the Event Horizon. New studies of galaxies showed high velocity stars spinning near the center of these galaxies, and x-ray emitting gasses traveling close to the speed of light, seemingly proving that Black Hole's existence (Bartusiak, pg. 210).

In his book, 'Hawking on the Big Bang and Black Holes', Stephen Hawking explains empty space as not truly empty, but full of virtual particle pairs consisting of both positive and negative energy. Hawking goes on to explain that as these particles begin to be pulled into the Black Hole's Event Horizon, the pairs split up; the positive particle being ejected out from the Black Hole as "thermal emission" while the negative particle "tunnels" uneventfully into the Black Hole (Hawking, 1994, Hawking on the Big Bang, pg.88).

Although Hawking does not subscribe to the "Aether Theory", he seems to indicate that "empty space" is not empty after all and that "local energy density" varies with varying amounts of these particles from place to place (Hawking, 1994, pg. 87). This strongly parallels the hypothesis that is being promoted by this book, which all matter consists of condensates of some material, which energy cannot exist on its own, and is simply manifested by the movement of this material.

Hawking also suggests that the positive energy particle is being ejected away from Black Hole as "thermal emission" or energy radiation of some sort. Could it be that Black Holes are simply made up of the predicted, Bose-Einstein material, which is created by mass so super cooled, and that it collapses into its self as one solid material? Super cooled due to the fact that matter has all the energy "wrung out of it" as radiant energy, before it falls into the Black Hole?

Theoretically, as mass approaches a Black Hole, energy literally is squeezed out of it; evidenced by the x-rays and the brightness of the center of galaxies such as MCG6-30-15. Here, it would not be the attraction of the super massive Black Hole that pulls in all material, but the huge differential in pressure between energetic particles in the universe along with cold condensed particles in the Black Hole. The universe pushes in, in order to create gravity, not the mass of the Black Hole that pulls in on the matter surrounding it. As energy is squeezed out of these particles, the particles move slower and slower until all energy subsides and unifies itself with the huge Bose-Einstein mass of the Black Hole.

The Black Hole is therefore "black" not because of its intense gravity that prevents light from escaping, but simply because it has no energy at all; as all energy is forced out of mass before it enters the Black Hole. Again, this suggests that the universe is cooling and contracting as energy bleeds away from its core. The expanding universe may again, simply be an illusion, and the Big Crunch is under way.

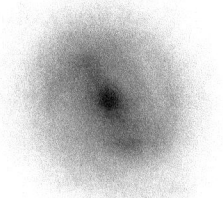

CHAPTER V

DISCUSSION

Review of Theory

Einstein and the other scientist at the turn of the 19th century were baffled by the fact that the speed of light was always the same velocity to an observer, no matter what the velocity of the observer or the velocity of the object emitting the light. Lorentz had shown that if an object was moving, that it contracted in the direction of its motion, increasing the distance light had to travel, therefore allowing the "perceived "speed of light to remain the same. Motion was warping space and time. This led Einstein to develop his theories of Relativity.

The Theory of Special Relativity explains that motion is relative; Einstein showed this with the train and embankment thought experiment. Here two bolts of lightning strike simultaneously in front and behind a moving train. There is one observer midpoint between the lightning strikes on the moving train and one observer midpoint the lightning strikes on the embankment beside the train. Although the lightning strikes occurred at exactly the same time to the observer on the ground, the observer on the moving train saw the light from the front of the train first because of his forward motion. By the time the light meets this observer, he has moved towards the lighting in front of the train and away from the lightning behind the train.

In this text, we have reviewed a similar thought experiment with a train blowing its whistles simultaneously from the engine and caboose. We saw the different times the whistle was heard by the two observers and also saw the Doppler Effect by the movement of the train. Here, movement or motion caused a different reality for both observers.

The Theory of General Relativity went on to show that mass also has an effect on time and space. Mass causes space to bend around it, which in turn causes mass to attract mass, the effects of gravity. This also is what causes light waves to bend and slow down, manifesting itself in gravitational lensing and red shift. The effects of mass also cause clocks to change their tempo, in effect changing time. Therefore both mass and velocity have a direct influence on time and space. The greater the mass, the more space and time are warped and the greater the velocity, the greater time and space are warped.

Einstein also showed that gravity and acceleration are equivalent with his Equivalence Principle. This was shown with a thought experiment where an observer was floating in space in a large room with no gravity. A force was applied to the room that caused it to accelerate at a constant rate, forcing the observer onto the floor. The observer, not being able to see outside, assumed that this force was gravity. This asserted that gravity and acceleration is essentially the same thing.

Another thought experiment was performed in this text in Blast Off. Here velocity was studied comparing the speed of one rocket stage to another in outer space, showing that velocity was relative to a reference point – one of the other stages. This point was also shown with the baseball being thrown from a moving vehicle earlier in this text. The point here given was to show that if speed was relative, the effects of speed would also have to be relative. This means that since velocity changes time, that this time change would be relative to different observers at different relative speeds.

As mass increases at speeds close to the speed of light, this mass change must also be relative to different observers. An object would therefore be getting larger to one observer than to another observer at a different relative velocity; how can this be? Just as in Einstein's train, this can only be a **perceived** difference and not reality. The only way that we can get to the point where these effects of velocity are real, is to create a stationary point in space in which all object's velocities are referenced to.

Review of Relativity

Let's look at this again. We all know the scenario where two cars are traveling toward each other; each car is traveling at 50 mph. That is 50 mph relative to the ground. Relative to each other they are traveling at 100 mph. As they pass, one passenger will throw a baseball to a passenger in the other car at 20 mph, relative to the car it is thrown from. Relative to the passenger catching the ball, say he had a radar detector aimed at the ball, the ball will be traveling at 120 mph. If an observer standing at the side of the road aimed a radar detector at the flying ball, it would read 70 mph (50 + 20). Of course here on earth we "usually" use the earth as the reference for speed. An aircraft on the other hand must use airspeed because the lift of the aircraft depends on it. In space, outside our solar system, what is our reference for velocity? Let's put a satellite up in space and launch a rocket from it at 80,000 miles per second (mps). Let us now launch a smaller rocket from this rocket at 80,000 mps. Then another rocket from this rocket is launched at 80,000 mps. The third rocket should be traveling at 240,000 miles per second in reference to satellite from which the original rocket was launched.

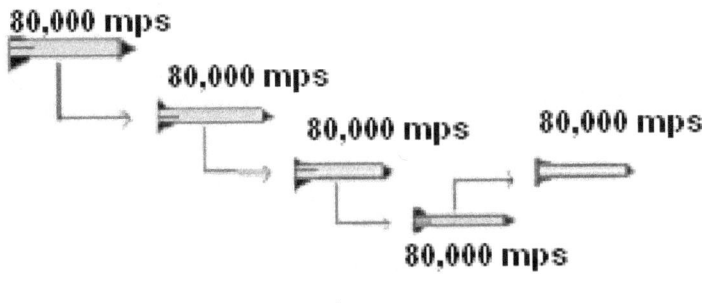

Rockets

Now what is the speed of light? 186,000 mps! What is it relative to? Electromagnetic energy (light, radio waves, x-rays) is always at the speed of light when observed, no matter what the position or velocity of the observer or receiver. There is no other reference point. The movement of any object is always measured by its change of position over a given period of time in reference to something; a change that may be in reference to its original position in space. Any object in space can be considered to be stationary unless it is changing in acceleration (speeding up or slowing down). In the illustration above, all objects are moving slower than the speed of light in reference to each other except, the first rocket launch sight (the satellite) relative to the last rocket. Neither the satellite nor the last rocket are accelerating or decelerating. The first may consider itself stationary or the last may consider itself stationary. The first rocket may be moving away from the last at a speed above the speed of light or the last rocket may be moving away from the first at a speed above the speed of light, which is it?

Einstein stated that an object's mass increases with its velocity. But what is its velocity relative to? He must be referring to stationary space! What is stationary space? How fast is our system moving relative to stationary space? If an object is moving at a high-speed relative stationary space, its mass increases. Does that mean that if an object slows down to zero velocity relative to stationary space, it will be non-existent and have no mass?

If there is no state as stationary space, and the last rocket moves away from the first at the speed of light, then the last rockets mass increases. If the last rockets mass increases, the first rockets mass becomes smaller relative to the last; or is the first moving away from last at the speed of light? If this is the case, then the first rocket should be getting larger, not smaller, in relation to the last rocket.

I'm afraid reality is not as complex as it may seem. Let's be careful not to make something that we perceive to exist, to be absolute. Remember the rocket and the clock? As the rocket moving away from us sends a radio signal of its clock, we perceive the clock to be ticking slower. As it returns, it is perceived to be faster. But reality is different. The clock never changed its rate of ticking. It never changed time!

This *medium*, the "aether", must exist. The whole electro-magnetic spectrum is dependent on it. This *medium* is mass and relatively stationary in undisturbed space. The smallest of all particles, it may consist of all three energy states, positive, negative and neutral. The neutral state may simply be a combination of both positive and negative states. The *medium* particles maintain an equidistant spacing from each other, depending on the density of the surrounding space. The *medium* is contained in both mass and empty space. Within mass and close to strong gravitational fields, the *medium* is much denser and will affect the transmission of electromagnetic waves through it. The velocity of electromagnetic propagation depends on the density of the *medium*. In materials that have high density, the *medium* is denser, such as in glass and plastics, and the velocity of the energy through them is slightly slower. The *medium* is an elastic solid, unlike the assumptions of Michelson and Morley, and is compressed by moving objects, gravitational, electric and magnetic fields. It is in fact the change in density and pressure of this medium that generates these fields. Outside of any movement and external fields, such as magnetic of gravitational fields, the *medium* falls to a stationary and neutral "pressure". In front of a moving object, the *medium* compresses and therefore increases in "pressure". In the wake, it falls to negative pressure and expands; the "pressure", being the ratio of the spacing of its particles in reference to stationary, neutral space. All electromagnetic energy is transferred through the *medium* from an emitter to a receiver. The number of particles in a given space determines the velocity of the electromagnetic waves propagation. The denser the *medium*, the less elastic the *medium* becomes and slower the electromagnetic propagation.

Stephen Hawking, in his book, A Brief History of Time, suggests that there may be a form of matter that is uniformly distributed, yet still undetected that could raise the average density of the universe (Hawking, 1988, pg. 48). Could this be the medium that bends light from stars and slows down radar signals bounced off planets? Could this be the dust that causes Mercury to precess more than expected in its orbit close to the sun? Is this the force that changes the time in the GPS satellites or the extra mass contained in the galaxies that prevents the outer stars from flying out into the open universe?

Gravity of the Matter

Gravity then in summary has been shown to be a geometric effect of curved Space - Time and is not a propagated source of energy as we see in electromagnetic and sound waves. The effects of gravity are therefore "local" and instantaneous and not in the form of Gravity Waves that propagate. This has been shown with the simple principle of aberration of sunlight, how that the perceived position of the sun in the sky is approximately 8.3 minutes behind its true position due to the transit delay in light waves through space. Gravitation has no such aberration or delay; research has shown that the gravitational position of the sun in reference to the earth in its orbit around the sun is the same as it true position. If this were not the case, the earth would orbit around the aberration and not the true position of the sun, eventually resulting in the decay of the orbit. If gravity was propagated in waves and traveled the great distances of space over time, orbits of bodies could not be maintained and the universe would break apart.

Another way to understand gravity is through the movement of the earth in orbit around the sun. It takes light 8.3 minutes to travel from the sun to the earth, therefore, where you see the sun in the sky is where the sun was about 8.3 minutes ago; in reality the sun's true position is about 20 arc seconds from its visible position. However, when measuring the acceleration vector for the Earth's motion in orbit around the sun and comparing the gravitational position of the sun to the apparent visual position due to transit delay, there is a difference. If gravity propagated through space at the speed of light, there would have been no difference. Gravity therefore, unlike the electromagnetic forces of light, radio waves, electromagnetic fields and magnetism, has no detectable aberration or delay. If gravity traveled at the speed of light, the earth would tend to rotate around the sun's apparent position and not its true position causing the earth's orbit to change slightly in distance on each rotation. Gravity therefore must propagate at much higher velocities than light, possibly even instantaneously in time and cannot be considered as a force of nature that propagates. Simply stated, gravity must purely be a geometric effect of curved space and time (Space-Time). The thought then comes to mind, if space is empty, what is being curved?

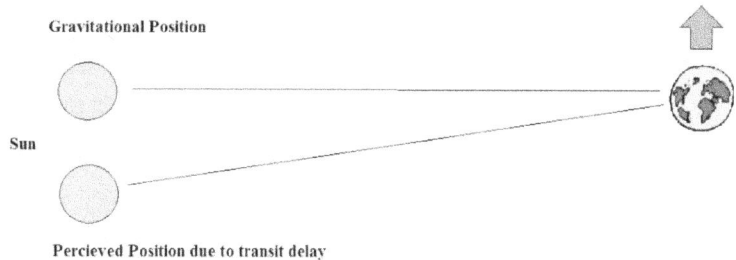

Geometric Effect

Static charges that attract and repel act in much the same way and do not propagate or take time to affect other objects. This seems to indicate that an invisible field surrounds and moves with all particles of mass and instantaneously causes what Einstein called a "spooky" action at a distance (Isaacson, pg. 458).

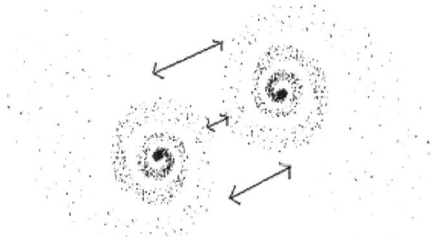

Like Charges Repel

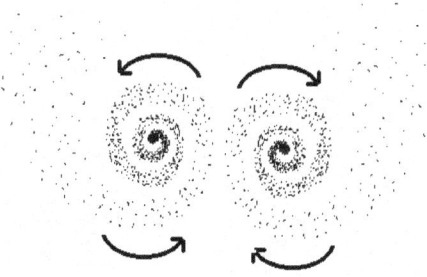

Opposite Charges Attract

Imagine in a thought experiment a steel girder being hoisted by a crane into position on a skyscraper with a line of dominoes on top of the girder. As one end of the steel girder impacts its positioning point, the dominoes begin to fall. Domino's fall at approximately 1 meter per second therefore taking about 50 seconds to travel the length of a 50 meter girder. Although the speed of the dominos takes only 50 seconds to travel the length of the beam, the beam itself stops its forward motion immediately. In much the same way, gravity like the girders motion occurs immediately whereas the light waves like the dominoes takes time to propagate over a distance.

Another analogy that can be used it that of helium filled balloons. A balloon begins to rise when it is released because the balloon is "lighter" than air. The Ideal Gas Law which was developed by Emile Clapeyron in the early 1800's. This law was established in order to explain the properties of gasses in relation to each other and is expressed as $pV = (w/mw) \times (RT)$ or simplified as $pV = nRT$. The p in the equation represents "pressure," V represents "volume" in "moles," w/mw represents weight (gravitational force) divided by molecular weight (the substance of the gas expressed as n), R is the gas constant and T represents the temperature of the gas.

This formula works for buoyancy of objects in all fluids such as a piece of wood or lead pipe in water. The Ideal Gas Law states that an object will "float" in a fluid due to the fact that the force exerted on the object by the fluid around it is greater than the force of gravity on that object. This formula takes gravity into account in order to calculate the buoyancy of an object and cannot be used to calculate gravity itself. However the same principle can be used; an object moves in a specific direction because it is less dense or denser than the fluid surrounding it. Therefore, if an object in a fluid has a fluid denser in one direction and less dense in the other, it will move toward the area which is less dense.

This very same principle can be seen in the weather formation of storm systems on earth, especially in the spin and force of a hurricane. The core of a low pressure area consists of air that is less dense that the air in a high pressure area. This creates a pressure differential that causes air to flow as "wind" in toward the low pressure areas; the differences in pressure ranges from low to high as pressures are measured out from the center of the system which can be charted as pressure gradients or isobaric lines. A sailboat on a lake will be blown and move toward the center of the low pressure system as the air flows from a denser region to a less dense region. The Gravity of the Matter can be better seen through this hypothesis.

ImGui

CHAPTER VI

CONCLUSIONS

Assumptions

The **first assumption** we will make is that energy is the movement of mass. Scientifically speaking, energy is a variable quantity that is a characteristic of an object (mass). There are several ways to describe energy, kinetic, potential, gravitational, electromagnetic, chemical and nuclear. Energy can be transformed from one form of energy to another such as light into heat, but cannot be created nor destroyed. Action at a distance, such as the transfer of energy as light or gravity must therefore be transferred through the movement of mass; this may involve the photon and the graviton as taught in modern physics.

Light moves in waves and clearly generates the Doppler effect when objects that give off energy move. Waves move through some sort of medium that is relatively stationary in space; an example is waves in water and sound through air or any other object such as steel. Here the **second assumption** is made that light waves must move through <u>something</u> that is relatively stationary in space.

Light waves are transmitted in both longitudinal and transverse waves. Longitudinal waves can move through any solid or fluid, however, transverse waves can only move through solids. The **third assumption** therefore states that the substance that light waves moves through must be a solid – generating the problem of how planets and stars can move in rigid space.

If space is rigid, then all movement must consist of energy waves passing from one particle of this semi rigid substance to another. Standing waves manifest themselves normally as interference patterns generated in two or more waves of similar harmonics, interacting with each other in a stationary medium. Our **fourth assumption** is that matter may be simply manifested as energy waves moving through a solid medium. A lenticular cloud forming over a mountain ridge may seem to be created of stationary particles. This cloud however is formed by moving moist air moving over a mountain top, condensing its moisture in the cold winds at high altitude as the air drops below the dew point. As the air moves down the downwind slope it warms up and evaporates. This constant movement of air creates a "standing wave", a cloud that seems to stand still, however generated by fast moving condensing then evaporating water particles.

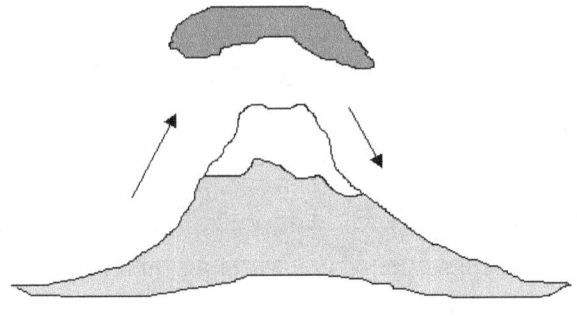

Lenticular Cloud

Another form of a standing wave can be seen as vapor trails over the wing of an airplane especially on a moist or rainy day. Here, similar to the lenticular cloud, moving air condenses in the low pressure region above the wing. Although this condensation seems to be stationary to the passengers in the aircraft, the air moving through this area may be moving at speeds over 500 miles per hour. Moisture in the moving air condenses momentarily in the low pressure area and then disappears as it moves away from the wing creating a standing wave.

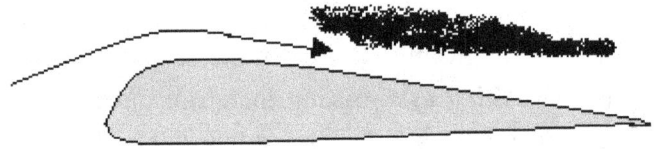

Low Pressure Condensation

Similarly, photons and electrons may simply be the manifestations of confined energy. Energy moving as waves at the quantum levels must be transferred in packets of Plank's Constant and cannot be divided. Energies at these quanta will then move in a straight line or possibly in a confined, circular orbit – as if trapped; it cannot be dispersed. This trapped or confined energy then manifests itself as a moving particle in a "stationary" medium.

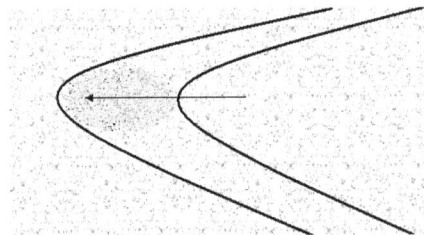

Photon Moving as a Particle in a Solid Medium

Energy of a photon may move from particle to particle in a stationary medium, yet confined to a specific size and area due to Planks Law. The particle moves in a straight line at the speed of light until absorbed by some other substance, normally an atom of some sort. A photon would therefore manifest itself as a moving particle even though it would simply be energy moving though a stationary solid substance.

An electron would likewise move through this medium as the photon does, however, it would only exist at specific regions known as shells around the atom. As it would possibly gain angular momentum, due to absorption of energy in the form of a photon, it will reappear at another shell further out. Energy absorbed above a specific level will eject this extra energy as a photon away from the atoms nucleus. An electron would therefore manifest itself as a moving particle even though it would simply be confined energy moving though a stationary solid medium around a condensed particle or group of particles – the atom.

It is possible however, that Keplerian dynamics and angular momentum do not pay a role as we may think with these particles. Since these particles may simply be the manifestation of energy, there is no true net movement of mass, just waves of energy moving through a solid substance. The velocity of the electron wave may therefore increase in speed the further away it moves from the center of mass or atom; as the medium becomes less dense and the closest electrons to the atom may move the slowest in the densest medium.

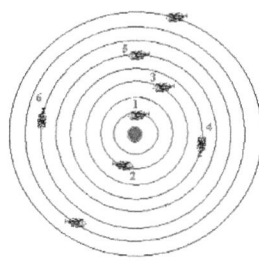

Electrons as Waves Moving Around a Nucleus

Without this stationary substance that light waves would move through, there would be no reference for the speed of light. Photons as moving particles would move relative to the object ejecting them or absorbing them, and light could not move at a constant velocity. Dual stars orbiting each other and moving in opposite directions as seen by Willem de Sitter, would give off light at different speeds to different observers. The mere fact that light generates Doppler shift is proof that light moves as a wave through a medium. Light or any other form of electromagnetic energy would therefore always move at a specific speed relative to the observer, similar to the speed of sound moving in the submarines sonar detectors dome or the speed of sound in the ear canal of someone hearing a trains whistle.

The **fifth assumption** will then state that other forms of mass would form from the cooling and coalescing of particles of this substance that is dispersed throughout the universe. This substance would contain hot and cold areas, cold areas having more densely spaced particles in this substance with a lower "pressure". This pressure differential of low pressure close to particles and high pressure away from particles will cause particles to move together – the generation of gravity; the more the mass, the higher the differential pressures pushing mass together.

We can visualize this process by a simple thought experiment. Imagine if you will a room separated in the middle by a thin plastic membrane, sealed off so that air cannot pass from one side of the room to the other. The room will be quite warm, say 38⁰C or 100⁰F and very humid. We then, using a heat exchanger with a refrigerant of some sort, consisting of a condenser and evaporator, transfer heat energy from one side of the room to the other with no net air movement; both sides air tight and insulated. Over time, one side of the room will become cooler and cooler as the other side of the room becomes hotter and hotter. Soon the cold side of the room begins to fog up as water vapor starts to condense in the air and droplets of water begin to form on cooler surfaces; it doesn't take long until frost begins to form on the coldest objects in the room. Now, take notice of the thin plastic membrane, how it now bulges into the cool area of the room. Here, the hot air, having higher pressure, pushes in on the cold area, ballooning if you will, the membrane into the cold room; the hot region, pushing in on the cold region.

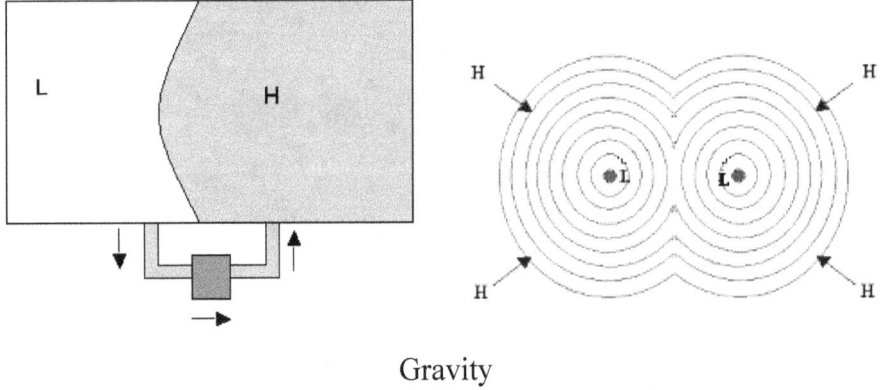

Gravity

Back to the real world, energy such as light waves moving through these denser regions, would move slower and refract, bending the light, similar to how light moving from air into water in a glass, makes a spoon seem to bend. This creates a **sixth assumption** that the speed of light varies according to the density of the medium it passes through, higher densities being close to mass.

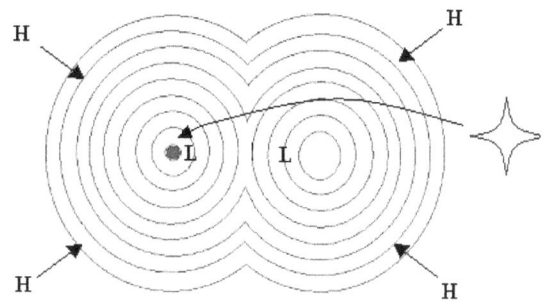

Starlight Bent by the Suns Gravity

The **seventh assumption** then states that if light moves slower close to large masses and that time itself will move slower. The opposite would then be true that the lack of gravity would in fact speed up time. Black holes, having the most mass therefore would have the largest effect on time to the point at the center; a singularity exists where time stands still. Time may vary from region to region, each area of space having its own "local time"; time however, could never be reversed, only stopped or sped up. Could these singularities be the portals to other dimensions?

Just as with sound in air or water, energy moving through this medium in space will move at a constant velocity depending on the "density" and "temperature" of the medium. Even though nothing that consists of energy waves could move faster than the speed of light, the speed of light will still vary from one region of the universe to another depending on the density of this medium. This then would suggest the **eighth assumption** that wormholes do exist. Wormholes would be areas in the universe almost void of this medium. Energy moving in this area could move at velocities thousands of times faster than the speed of light in "near space" or space close to massive objects such as our sun. If mass merely consists of energy waves moving through this medium, then mass could very well be able to move through these areas at very high speeds, allowing one to "pop out" on the other side of the universe.

If this medium is dispersed throughout the universe in all directions and was created by the "Big Bang", it then should become less dense, the further away from the center of the source of the energy, to the point where it no longer exists. At this point energy can no longer propagate (similar to how sound waves cannot propagate out into space) and matter and time can no longer exist – the edge of the universe, the **ninth assumption**.

The **tenth assumption** is that this universe is slowly collapsing. Edwin Hubble's discovery in 1929 suggested that galactic red shift in light from stars and galaxies increases in proportion to their distance from earth. His studies showed that 38 of 40 galaxies observed are rapidly moving away from each other, an omni-directional galactic expansion. Galaxies furthest away are moving the fastest – some at 57,000 kilometers per second or 13 million miles per hour – 2% the speed of light. The problem here is that this is evidently omni-directional – indicating that we are the center of the universe. If we were viewing this expansion out on the edge, only certain areas focused in one region would have this red shift. It must be that either we are the "center of the universe" or this observed redshift, therefore, is simply an illusion; an illusion due to the shrinking of the universe around us as it cools, loosing energy out into the vastness of the universe.

Since it takes time for the universe to cool and also takes time for light to reach us on earth, the changes to the universe will be seen uniformly in all directions and be relatively consistent at various distances. This however suggests a possible "Big Crunch" as energies subside and particles coalesce. Eventually as the medium between all objects condenses and cools, there will be nothing to keep objects apart; this expansion medium will no longer exist. Higher energies outside will eventually push all matter together into one place. Here atoms can no longer exist, fusing together into what has been called the **Bose–Einstein Condensate** or BEC's, huge molecules that form at near 0 Kelvin or absolute zero and possibly what Black Holes are made of.

CHAPTER VII

HYPOTHESES

In summary, our **first** hypothesis is that energy cannot exist without mass, much like velocity cannot exist without the bullet. There can be no such thing as pure energy; mass cannot become energy and energy cannot become mass. Energy is simply manifested as the movement of mass in some way or form.

Second, we hypothesize that light waves must move through some sort of medium, much like waves move through water or air. In order to have a wave, some sort of medium or field must be waving; something like what was once called the aether. The study of the movement of galaxies suggests that something else exists within these galaxies that tie all the stars in them together, causing them to move in unison counter to Keplerian dynamics. It is now believed that Dark Matter is this component that ties the stars together within the galaxies. Dark Energy is also now theorized in order to support Hubble's Law and the apparent expansion of our universe and the force that pushes galaxies apart. Today it is believed that the universe is composed of 73% of Dark Energy, 23% of Dark Matter, and 4% ordinary Baryonic matter of which we are made of. Could this Dark Matter and Dark Energy be the manifestation of this medium?

On July 4, 2012, the Higg's Boson, also known as the "God particle" was discovered at the Large Hadron Collider at CERN (Conseil Européen pour la Recherche Nucléaire) near Geneva Switzerland. Although much more complicated then said here, this particle was only theorized to exist for some forty years. It is a particle that has no spin, no electric charge and exists in "empty space" as the Higgs Field and is believed to be the mechanism in which particles acquire mass. It was the missing link in the physics Standard Model that was needed to explain how the interactions of the electromagnetic, weak, and strong nuclear forces work. Could this be the medium in which matter is hypothesized to be manifested in through the energies of Standing Waves?

The **third** hypothesis is that, whatever light or electromagnetic energy is moving through (the medium or aether as it might be called) must be considered to be an elastic-solid. This is due to the fact that transverse waves, waves that are polarized, can only move through solids and cannot move through fluids.

The **fourth** hypothesis is that mass consists of standing waves within this elastic-solid medium. This is why the planets, stars and galaxies are not rigidly in place and move freely within this medium. Picture the universe, if you will, as a three-dimensional television screen where individual pixels close up do not move. Simply speaking, the energy that is changing the pixels color and brightness; changes in amplitude move from pixel to pixel, giving the illusion of movement while the pixels remain solidly in place.

The **fifth** hypothesis is that particles of the subatomic world are this illusion; energy moving through this medium created by the cooling and coalescing of these particles or confined pressure waves moving through it. This is why the Michelson Morley experiment failed; mass is the medium, manifested by energy moving through it. Pressure differentials in this medium create the illusive condition called gravity. Here, higher energies away from mass press in on the lower energies of colder, condensed mass (including the fiery stars), this causes mass to attract mass.

Our **sixth** hypothesis states that closer the medium is to the coalescing mass, the denser the medium becomes, causing light passing through this area to change velocity. Since light moves in waves, refraction occurs, causing light wave to bend as they pass close to large masses.

Our **seventh** hypothesis is that time will vary with gravitation, creating what is called "local time". Time, is the change in position of energy moving through the medium, frame by frame. Energy, however, moves slower closer to mass due to the mediums increased density. Time therefore will move slower the closer to mass an object sensitive to time comes. Time also will vary during acceleration and deceleration due to the "principle of equivalence" where gravity and acceleration are indistinguishable; it is not velocity that changes time, it is the acceleration or deceleration to that velocity of the object that changes "local time".

Our **eighth** hypothesis is that wormholes do exist. In the far reaches of the universe where the mediums density is almost void, energy may move at velocities much higher than our calculated speed of light. Since mass is the movement of energy through this medium, mass in this region may simply exist, then not exist, and then exist again, "jumping" as it were, far from where it came from. This jump is almost identical to the jump the electron encounters as it jumps from one shell to another, existing, then not existing, then existing again.

This brings us to our ninth hypothesis that at some point far away from the center of the "Big Bang" a place exists where space is completely void of this medium that mass can no longer exist. If mass is manifested as standing waves through this medium, mass and therefore also time cannot exist outside of this medium, just as sound waves cannot exist in space. This may not only occur just at the edge of Time-Space but may also come into play within the vast areas between the galaxies in currents of nothingness. Mass as it would enter these areas would simply "jump" across this emptiness as if entering a wormhole.

Small wormholes may also be created here at home simulating this "empty space" with energy, allowing one to "beam" from one place to another. This could be done by "energizing" or heating this medium in close vicinity to a transport vehicle to create this "empty space" much like how we heat gasses to accelerate rockets today. Energizing the medium would create pressure differentials in the medium similar to gravitational forces explained earlier. Since the medium that is being energized or heated, consists of so little mass, it may possibly take very little energy in order to create this force; much like the energy causing a balloon to float. It may simply be a matter of frequency, similar to a microwave oven heating your dinner. Using some sort of heat exchanger where cooling the medium in front of a transport vehicle and taking that energy to heat the medium behind the vehicle could induce this transport. Another possible form of propulsion could be inducing an imbalance in pressure magnetically with some sort of flux flow. This technology may allow the craft to hover or move at extreme velocities without the negative effects of g forces on its occupants, as if "falling" toward its intended destination, possibly falling into the sky.

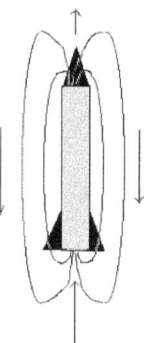

Our **tenth** hypothesis is that the universe is slowly collapsing. Redshift seen from galaxies furthest away from us suggests that rather than having an expanding universe, the universe is slowly collapsing. As energy in our universes core is radiated out into the outer reaches of space, energy is being lost and our core is cooling and shrinking. As we look back into time, an illusion is created that suggests that these galaxies are racing away from us omni directionally; in fact, it could be that we are racing away from them, as if falling into a "black hole." Black holes are all around us, the centers of probably all galaxies. Could it be that black holes are regions where the universe is gone completely cold? Where energy in mass has been completely "wrung out" and the "**Bose Einstein Condensate**" is created? Is this the beginning of the "Big Crunch"?

The **eleventh** and final hypothesis is that the speed of light is not a constant but a variable much like the speed of sound that is a constant under specific conditions. Sound travels in air at 761.2 miles an hour at 59 degrees Fahrenheit at sea level or 29.92 inches of mercury. In water, glass, steel or any other medium, the speed of sound will vary according to the medium it is traveling through and the temperature. Sound in glass is 10,155 miles an hour and through steel, 11,415 miles an hour.

We know that light travels through air at approximately 185,226 miles per second, or 99% the speed of light in a vacuum. Light passing through glass however, only travels about 114, 989 miles per second, or 62% the speed of light. In out eighth hypothesis we suggested that in the far reaches of the universe, the mediums density may be almost void and energy may move at velocities much higher than our calculated speed of light. All of our testing and calculations have been with tests here on earth and within our solar system on the solar plane between planets.

Solar and Galactic Plane

Could it be that within these gravitational fields of planets and the Sun and all the asteroids and dust within this plane that what we have calculated as the speed of light is wrong and that outside of the solar plane and galactic plane, space is far less dense and the speed of light is much higher?

REFERENCES

Allen, Tim (1996) I'm Not Really Here, Hyperion, 114 Fifth Avenue, New York, NY

Arago, François (1858) Oeuvres Complètes, 7th Edition, Volume 4, Paris

Aristotle (350 B.C.E) De Caelo, Book I, Chapter 2

Ashtekar, Abhay (2005) One Hundred Years of Relativity, World Scientific Publishing, 5 Toh Tuck Link, Singapore

Asimov, Isaac (1991) Atom, Journey Across the Subatomic Cosmos, Truman Talley Books, First Plume Printing, New York, NY

Barbour, Julian B. (2001) End of Time: The Next Revolution in Physics, Oxford University Press, 198 Madison Avenue New York, NY

Bartusiak, Marcia (2000) Einstein's Unfinished Symphony, Listening to the Sounds of $Spac^2e$ -Time, Berkley Publishing Group, New York, NY

Bless, Robert C. (1996) Discovering the Cosmos, University Science Books, 55D Five Gates Rd, Sausalito CA

Bord, Donald J. (2000) Inquiry Into Physics, Brooks / Cole Fourth Edition.

Borel, Armand (2001) Oeuvres, Collected Papers, Vol. IV, Springer Scientific and Business Media, 1150 65th Street, Emeryville, CA

Born, Max (1962) Einstein's Theory of Relativity, Dover Publications Inc. New York, NY

Brown, Julian (2000) Minds, Machines, and the Multiverse, The Quest for the Quantum Computer, Simon and Schuster, New York, N.Y.

Collins, Harry (1998) The Golem: What You Should Know about Science, Cambridge University Press, Cambridge CB2 2RU, UK

Dekel, Avishai (1999) Formation of Structure in the Universe, Cambridge University Press, Cambridge CB2 2RU, UK

Dennis, Mark F. (2009) The Speed of Light, Reviewing the history of "c", Golden Iris Publications

Dennis, Mark F. (2009) Traveling at the Speed of Light, Golden Iris Publications

Ditchburn, R.W. (1991) Light, Dover Publications, Inc., 31 E.2nd Street, Mineola, NY

Einstein, Albert, (1952) Relativity by Einstein, Special and General Theory, Tess Press, Fifth Edition

Evans, Alvis, J. (1985) Basic Electronics Technology, Tandy Corporation, Ft. Worth, Texas

Fresnel, A. (1818) Lettre d'Augustin Fresnel à François Arago Annales de chimie et de physique. 9, pg. 57–66

Galileo, Galilei (1632) Dialogue Concerning the Two Chief World Systems, Second Edition, University of California Press, Berkeley, CA

Glendenning, Norman K. (2007) Our Place in the Universe, World Scientific Publishing, 27 Warren Street, Hackensack, NJ

Hawking, Stephen (1988) A Brief History of Time, Bantam Books, Random House Inc. New York, N.Y.

Hawking, Stephen (1993) Hawking on the Big Bang and Black Holes, World Scientific Publishing, P.O. Box 128, Farrer Rd., Singapore

Hobson, Michael pg. (2006) General relativity: An Introduction for Physicists, Cambridge University Press, Cambridge University Press, Cambridge CB2 2RU, UK

Isaacson, Walter (2007) Einstein, His Life and Universe, Simon and Schuster Paperbacks, New York, NY

Kafatos, Menas (1990) The Conscious Universe, Springer-Verlag, New York, NY

Lorentz, H.A, (1909) The Theory of Electrons, B.G. Teubner, G. E. Stechert, Booksellers, New York, 129- 133 West 20th St. New York, NY

Newton, Sir Isaac, (1730) Opticks, a Treatise of the Reflections, Refractions, Inflections and Colors of Light, Fourth Edition.

Pitaeskii, Lev (2003) Bose-Einstein Condensation, Oxford University Press, USA, 198 Madison Avenue New York, NY

Poincaré, H. (1904) The Present and the Future of Mathematical Physics, Address Delivered to the International Congress of Arts and Science, St Louis.

Polkinghorne, J.C. (1984) The Quantum World, Princeton Science Library, Princeton University Press, Princeton NJ

Psillos, Stathis (1999) Scientific Realism, Routledge, 29 W. 35th Street, New York, NY

Sokolov, G., (1999) The Theory of Relativity and Physical Reality

Tippler, Paul (2003) Physics for Scientists and Engineers, Standard Version, W. H. Freeman Company, 41 Madison Avenue, New York, NY

Veatch, Henry (1978) Electric Circuit Action, Science Research Associates, Chicago Ill.

Van der Kamp, Walter (1988) De Labore Solis, Airy's Failure Reconsidered, Anchor Book & Printing Centre, Surrey, B.C., Canada

Wolf, Fred, (1989), Taking the Quantum Leap, The New Physics for the Non-Scientists, Harper and Row Publishers, New York, NY

APPENDICES

I cannot but regard the aether, which can be the seat of an electromagnetic field with its energy and its vibrations, as endowed with a certain degree of substantiality, however different it may be from all ordinary matter.

(Lorentz, 1906)

(Borel, Armand, pg. 690)

Indeed one of the most important of our fundamental aether not only occupies all space between molecules, atoms, or electrons, but that it pervades all these particles. We shall add the hypothesis that, though the particles may move, the aether always remains at rest.

(Lorentz, **1906)**

(Lorentz, H.A, pg. 11)

Lorentz proclaimed the very radical thesis, which had never before been asserted with such definiteness: the aether at rest in absolute space.

In principle this identifies the aether with absolute space. Absolute space is no vacuum, but something with definite properties whose state is described with the help of two directed quantities, the electrical field E and the magnetic field H, and as such, it is called the ether.

(Born, **1924)**

(Born , pg. 204)

BIBLIOGRAPHY

Allen, Tim. *I'm Not Really Here*, New York, Hyperion, 1996

Arago, François. *Oeuvres Complètes*, Paris, 1858

Aristotle. *De Caelo*, 350 BC

Ashtekar, Abhay. *One Hundred Years of Relativity*, World Scientific Publishing, Singapore, 2005

Asimov, Isaac. *Atom, Journey Across the Subatomic Cosmos*, Truman Talley Books, New York, 1991

Barbour, Julian B. *End of Time: The Next Revolution in Physics*, Oxford University Press, New York, 2001

Bartusiak, Marcia. *Einstein's Unfinished Symphony, Listening to the Sounds of Spac²e - Time*, Berkley Publishing Group, New York, 2000

Bless, Robert C. *Discovering the Cosmos*, University Science Books, Sausalito CA 1996

Bord, Donald J. *Inquiry Into Physics*, Brooks / Cole Fourth Edition, 2000

Borel, Armand. *Oeuvres, Collected Papers*, Springer Scientific and Business Media, Emeryville, CA 2001

Born, Max. *Einstein's Theory of Relativity*, Dover Publications Inc. New York, 1962

Brown, Julian. *Minds, Machines, and the Multiverse*, Simon and Schuster, New York, 2000

Collins, Harry. *The Golem: What You Should Know about Science*, Cambridge University Press, Cambridge UK, 1998

Dekel, Avishai. *Formation of Structure in the Universe*, Cambridge University Press, Cambridge, UK, 1999

Dennis, Mark F. *The Speed of Light, Reviewing the history of "c"*, Golden Iris Publications, 2009

Dennis, Mark F. *Traveling at the Speed of Light*, Golden Iris Publications, 2009

Ditchburn, R.W. *Light*, Dover Publications, Inc., Mineola, NY, 1991

Einstein, Albert. *Relativity by Einstein, Special and General Theory*, Tess Press, 1952

Evans, Alvis, J. *Basic Electronics Technology*, Tandy Corporation, Ft. Worth, 1985

Fresnel, A. *Lettre d'Augustin Fresnel à François Arago Annales*, 1818

Galileo, Galilei. *Dialogue Concerning the Two Chief World Systems*, University of California Press, Berkeley, CA 1632

Glendenning, Norman K. *Our Place in the Universe*, World Scientific Publishing, Hackensack, NJ, 2007

Hawking, Stephen. *A Brief History of Time*, Random House Inc., New York, 1988

Hawking, Stephen. *Hawking on the Big Bang and Black Holes*, World Scientific Publishing, Singapore, 1993

Hobson, Michael pg. *General Relativity: An Introduction for Physicists*, Cambridge University Press, Cambridge, UK, 2006

Isaacson, Walter Einstein. *His Life and Universe*, Simon and Schuster Paperbacks, New York, 2007

Kafatos, Menas. *The Conscious Universe*, Springer-Verlag, New York, 1990

Lorentz, H.A. *The Theory of Electrons*, B.G. Teubner, G. E. Stechert, Booksellers, New York, 1909

Newton, Sir Isaac. *Opticks, a Treatise of the Reflections, Refractions, Inflections and Colors of Light*, 1730

Pitaeskii, Lev. *Bose-Einstein Condensation*, Oxford University Press, New York, 2003

Poincaré, H. *The Present and the Future of Mathematical Physics*, Address Delivered to the International Congress of Arts and Science, St Louis. 1904

Polkinghorne, J.C. *The Quantum World*, Princeton Science Library, Princeton University Press, Princeton NJ, 1984

Psillos, Stathis. *Scientific Realism*, Routledge, New York, 1999

Sokolov G. *The Theory of Relativity and Physical Reality*, General Science Journal, 1999

Tippler, Paul. *Physics for Scientists and Engineers, Standard Version,* W. H. Freeman Company, New York, 2003

Veatch, Henry. *Electric Circuit Action*, Science Research Associates, Chicago Ill., 1978

Van der Kamp, Walter De Labore Solis. *Airy's Failure Reconsidered*, Anchor Book & Printing Centre, Surrey, B.C., Canada 1988

Wolf, Fred. *Taking the Quantum Leap*, The New Physics for the Non-Scientists, Harper and Row Publishers, New York, 1989

www.ingramcontent.com/pod-product-compliance
Lightning Source LLC
Chambersburg PA
CBHW080806180526
45168CB00006B/2345